Student Reg No:

10220108

Validation Number

By purchasing this training manual, you have paid for the course **training manual** as well as for your **course certificate** and **competency card**.

In order to qualify for the certificate and competency card, you need to submit the validation number below to your course tutor. The course tutor will input the number on your Candidate Result Form to ensure that you receive the course certificate and competency card that you have paid for.

It is very important that you do not lose this number. BPEC cannot issue certificates for forms that do not include the appropriate validation number.

All validation numbers are unique. Once this number has been used it will no longer be valid.

CW01590115

YOUR UNIQUE
VALIDATION NUMBER IS:

10220108

Attach this sticker on candidate's result form

Contents

Introduction

Contents

Introduction to the course

This course has been developed to enable plumbing and heating engineers to demonstrate their competence in the:

- Design, installation, service and maintenance of vented and unvented hot water systems;
- Understanding of vented and unvented hot water system design and installation requirements as they relate to building and water regulations

You will also cover methods of temperature control in open and unvented hot water systems.

Because of the changes to Building Regulations and the implications to vented systems, the course now includes vented systems as well as unvented systems. From now on the reference in the text will simply be 'hot water storage systems'.

The course is aimed primarily at practising plumbing and heating engineers who have an understanding of the fundamental requirements of hot and cold water systems and central heating systems or students who are nearing completion of a plumbing and heating vocational qualification at Level 3.

This training pack supporting the course serves:

- Firstly as a means of learning during the course, and
- Secondly as a reference document for use after the course

Course aims

At the end of the course the learner will be able to:

- Understand the requirements of Section G3 of the Building Regulations relating to the design, installation, service and maintenance hot water systems and how to comply with the regulations
- Understand the requirements of the Water Regulations/Bylaws in relation to hot water systems
- Identify typical faults associated with the installation or maintenance of hot water systems
- Understand the requirements of Part L1 of the Building Regulations detailing Conservation of Fuel and Power as they relate to the design, installation and service and maintenance of hot water systems
- Understand the application of Part P of the Building Regulations in relation to the installation of hot water systems
- Undertake a range of assessment tests to confirm their competence in the installation, service and maintenance of hot water systems.

Course pack

The course pack contains this introduction followed by 4 learning modules. Details of the modules are as follows:

Module 1: Vented and unvented hot water systems and their operating principles

Module 2: Design and installation of unvented hot water systems

Module 3: Maintenance of vented and unvented hot water systems

Module 4: Other applicable Building Regulations

Course structure

This course is designed to keep the time spent away from the workplace to a minimum. It is achieved through:

- A self study element, more about self study shortly
- Attending an approved BPEC centre where your tutor will:
  - reinforce the self study elements of the course and make sure that you understand the key points of the course and ensure you are updated on changes in regulations and technology
  - present a practical demonstration of the key components of a 'live system and it's operating principles' including installation and commissioning requirements
  - administer a series of assessment tests.

About self study

Self study is an extremely popular method of leaning and is often used in support of college or centre training courses. It enables some leaning to take place at home, thus saving time and costs.

Work through the modules contained in this pack in your own time and at your own pace. We recommend that study is undertaken in a location free from distraction.

The amount of study you undertake at any one time is up to you; however, we suggest that you read through each module several times until you thoroughly understand the content. Don't overdo it, try and do the learning in 'bite-size' chunks of no more one and half to two hours.

Once you feel comfortable with a module, try the questions contained at the end of the module without referring to the previous pages. Attempting to answer the questions will assist you in identifying whether you are beginning to understand the subject you are studying. The answers to the questions are detailed at the end of each module.

If you don't understand any part of the course after studying the modules, don't panic, just make a note of the subjects you're not sure of and raise them with your tutor when you attend the course.

The BPEC course

When you attend the course, your tutor will take you through the key knowledge requirements based on the course manual, so it's important at this point that you raise any issues which you don't understand or that you feel require further clarification

By the end of this session you should have attained sufficient supporting knowledge to cover the requirements of hot water systems.

On completion of the knowledge element of the course, your tutor will demonstrate the key operating principles of a live system and take you through the essential installation and commissioning requirements.

Module 1

Vented and unvented hot water systems and their operating principles

Contents

Aim of the module

On completion of this module (and following attendance on the BPEC Course) you should be able to:

- Identify the key method of operation of hot water systems
- State the differences between vented and unvented systems in terms of the key components and working principles of the system
- List the relative advantages and disadvantages of vented and unvented hot water systems
- Identify the legislation covering the installation and maintenance of hot water systems in the UK and its key requirements
- State the function of key components that are installed in hot water systems

Image 1 Unvented cylinder

This course is designed to enable you to demonstrate competence in the installation and maintenance of hot water systems in accordance with building and water regulations.

The supply of hot water in domestic dwellings was first introduced into the majority of households around 1920 to 1930. From these early, relatively simple applications, emerging technology has vastly improved the design principles and use of components and appliances in today's modern domestic hot water systems.

Early vented systems were referred to as 'direct systems'. This means that water heated in the boiler is fed directly into hot water storage vessel and ultimately is the water that is distributed to the hot taps.

Direct systems could also be heated solely by an immersion heater rather than by a boiler.

The introduction of central heating brought about the use of indirect systems. Here the water from the heat source, which also supplies the radiators, is kept separate from the domestic supply to the taps by the use of a copper coil (known as the heat exchanger) inside the cylinder.

The stored water is heated 'indirectly' via the heat exchanger. Single feed self priming cylinders were an alternative to the use of a coil type heat exchanger, but are no longer used in new installations.

During the earlier decades of the Twentieth Century mains fed hot water supply systems (unvented systems) were utilised throughout the industrialised world with the exception of the United Kingdom.

With the introduction of the Building Regulations 1985, a relaxation of the Water Byelaws current at the time and the Water Byelaws which came into force on the 1 January 1989, such systems were allowed to be installed in the UK.

Since then a number of amendments have been made to the Approved Document G to the Building Regulations and all unvented hot water storage systems must now comply with section G3 in England and Wales and the Water Regulations 1999. In Northern Ireland it is Regulation P5 and the Water Regulations 2009 and Technical Standard P3 and the Water Supply Byelaws 2014 for Scotland.

The installation of unvented hot water systems in the UK is now relatively commonplace with the unvented system providing a number of key benefits to the system user.

Although reference is made here to Approved Document G of the Building Regulations in England and Wales, the requirements are the same for the devolved administrations and the table below shows which parts of the relevant legislation in Scotland and Northern Ireland affect hot water storage systems.

England and Wales	Scotland	Northern Ireland
Part A	Section 1 (Structure)	Technical Booklet D
Part B	Section 2 (Fire)	Technical Booklet E
Part C	Section 3 (Environment)	Technical Booklet C
Part E	Section 5 (Noise)	Technical Booklet G
Part G	Section 4 (Safety)	Technical Booklet P
Part L	Section 6 (Energy)	Technical Booklet F1
Part P	Section 4 (Safety)	Part P

Unvented systems are a perfectly safe method of storing and supplying domestic hot water provided that the installation and maintenance of these systems is undertaken by properly trained and qualified individuals. To that end legislation is in place which identifies who may carry out work on these types of system.

Let's read on and look at hot water systems in a little more detail!

Vented and unvented hot water systems

Firstly, let's take a look at an illustration of an early vented system and its key components:

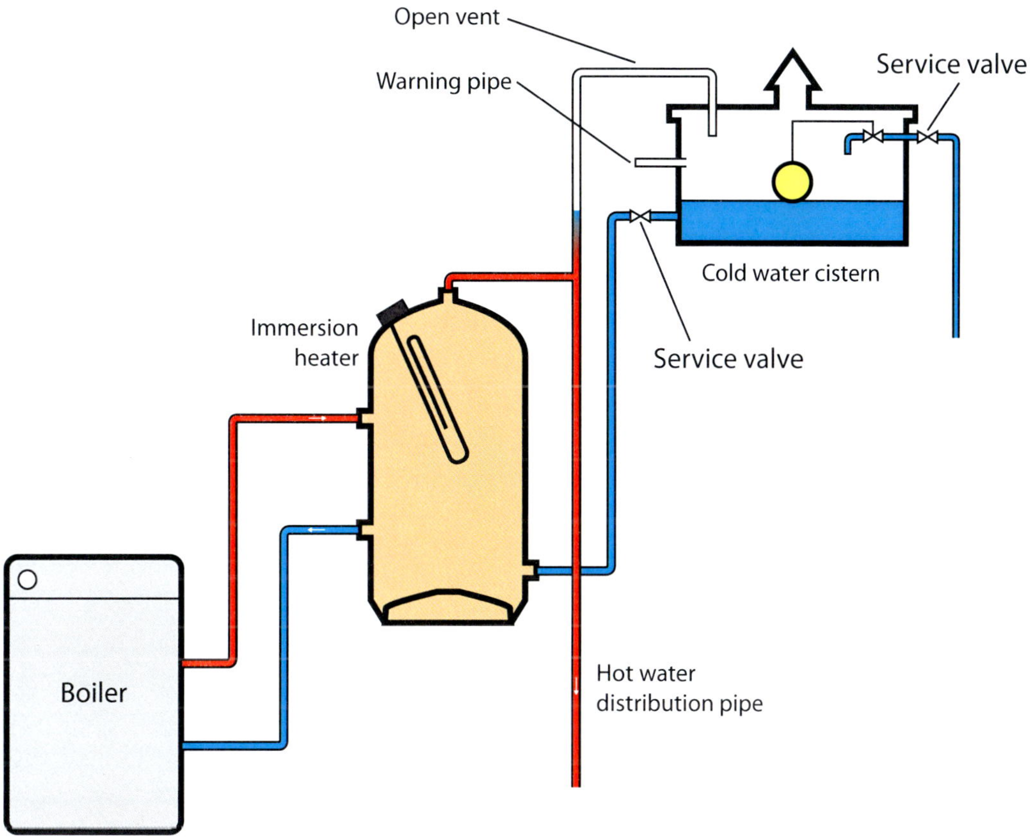

Image 2 The direct vented hot water storage system

The supply of hot water to houses within the UK has traditionally been provided from a cold water storage cistern usually located in the roof space, which then feeds a hot water store cylinder.

The rate of discharge and water pressure at the tap is dependent on the height or head of water between the level in the cold water cistern and the draw off tap.

Because water expands as it is heated, the volume of water in the system has to be accommodated. In a traditional vented system, with a cold water storage cistern, the expansion takes place primarily back up through the cold feed pipe and into the cold water storage cistern; thus the volume of water increases in the cistern and the water level rises. There is no pressure increase in the system because the cistern is open to atmosphere.

Finally, the traditional vented system is provided with a key safety measure; the open vent pipe (or often incorrectly called expansion pipe).

These systems were unhygienic as many had no lid on the cold water cistern so contaminants such as birds and mice could get into the water. The water in the cylinder circulated through the boiler and was then drawn off at the taps.

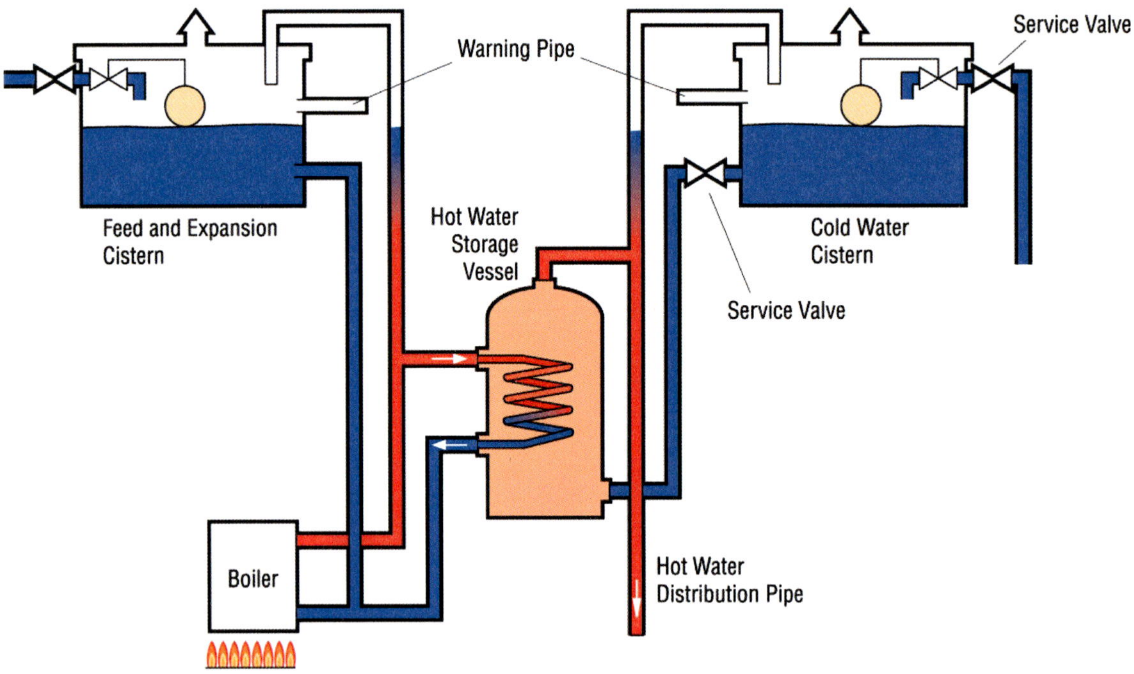

Image 3 The indirect vented hot water storage system

Here are a few important factors about double feed indirect hot water systems:

- The system requires an open vent and cold feed pipe connecting into the primary pipe circuit, or it can be separately fed into the boiler.
- The system requires a separate feed and expansion cistern.
- The heating appliance (boiler) may have a cast iron heat exchanger.
- No type of shut off valve should be fitted to the open vent pipe.
- If the open vent pipe is not connected to the highest point in the primary circuit, then an air release valve must be fitted, preferably an automatic one.

Note: This is an illustration and not a technical installation diagram

Use of immersion heaters in direct and indirect hot water storage systems

Some existing systems are heated solely by the use of immersion heaters. Due to a number of high profile scalding incidents which have resulted in fatalities, any new immersion heater must now be compliant with the requirements of BS EN 60335-2-74.

This standard requires the heater to contain:

- An operating thermostat which can be adjusted to user requirements (normally 60-65°C).
- An overheat thermostat which is set by manufacturer to operate at a temperature of typically 80°C. The overheat thermostat is of the non self-resetting type meaning that when it activates it must be re-set by the plumber after the fault has been investigated.

The purpose of the overheat thermostat is to activate and prevent further heating of the cylinder if the normal operating thermostat fails.

The following Image 4a shows an example of an immersion heater thermostat which is sited inside the cover of the heater. It can be seen from the photograph that the thermostat assembly includes an overheat thermostat by the inclusion of a reset button. Image 4b shows an immersion heater thermostat of the type which is built into the cylinder, it too has a reset button which needs a cover removed to access.

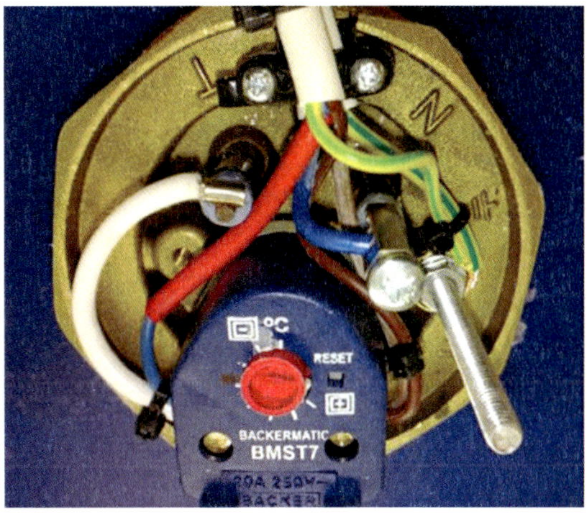

Image 4a **Rod type thermostat**

Image 4b **Built-in thermostat**

It is now recommended that when carrying out repair/maintenance on an existing cylinder containing an immersion heater with single operating thermostat that this be replaced with a dual thermostat in order to prevent any future high risk scalding incident.

The need for the open vent pipe

Quite simply when water is heated it expands or increases it's volume; for temperatures up to 100°C the expansion of the water or increase in its volume is usually up to 4% of the system contents at cold start up.

When water is heated above 100°C it expands so much it becomes water vapour in the form of steam as you will know from your kettle!

Here's an example of temperature rise and increase in volume:

If a system contains 200 litres of water at 10°C cold start up and is then heated to an operating temperature of 100°C, the system volume will increase to approximately:

200 litres (original volume) to 200 litres x 0.04 (expanded volume)

200 + 8 = 208 litres an increase of 8 litres

In normal operation a rise of 50°C (from 10°C to 60°C) will result in an increase in volume of 4 litres but the system needs to allow for containing the increase from fault conditions.

So, we need to make provision in the system for that water to expand. Primarily this is into the cistern via the feed pipe, or to the atmosphere through the open vent pipe. Without such provision, the expansion of the water or increase in its volume would increase the pressure inside the system.

The open vent pipe must be of an appropriate size (not less than 19mm internal diameter) and must connect from the top of the hot water storage vessel to a point open to the atmosphere above and over the level of the water in the cold water storage cistern. This means the open vent will pass through a rubber grommet in the lid of the cold water storage cistern and terminate above the maximum water level of the cistern.

In addition to the vent pipe and any thermostat which is used to control the temperature of stored hot water, hot water storage systems should also incorporate either:

- For all direct heat sources:
  - Non-self re-setting energy cut out to disconnect the supply of heat to the storage vessel in the event of the storage system overheating
- For all indirect heat sources:
  - An overheat cut out to disconnect the supply of heat to the storage vessel in the event of the stored water overheating so that the temperature of stored water does not exceed 100°C

  OR
- An appropriate safety device e.g. a temperature relief valve or a combined temperature and pressure relief valve to safely discharge the water in the case of significant over heating

Closed vessels

Remember; if water is contained in a **closed** vessel (unvented cylinder) and is subjected to heating, the water in the vessel will expand. If there is no room for expansion the pressure inside the vessel begins to build up. Let's explore what would happen if we did not allow for expansion of water in the following example.

If we go back to our previous calculation and the 8 litre increase in volume. 8 litres does not sound a lot of water or potential pressure build up and that is true, however, in undertaking the calculations we have assumed that the water temperature will always remain below 100°C.

In reality, in most storage systems water is stored at 60-65°C so the actual expansion in volume is more like 2.5% or 5 litres for our example.

However, in practice, there may be situations, due to a fault for example, where the temperature of the water exceeds 65°C, so for this reason systems are always sized for the maximum expansion of 4% plus an allowance for safety.

High performance storage cylinders

Cylinders of this type offer energy efficiency and lower fuel running costs. They give ultra fast recovery times for hot water, and this improved efficiency and performance allows for a reduction in both the size of the cylinder and its storage capacity.

The cylinder is designed for use on indirect fully pumped hot water systems, incorporating a motorised zone valve and thermostat temperature control.

Unlike conventional indirect copper cylinders which utilise a single internal 28mm diameter coil heat exchanger, high performance cylinders are fitted with a multi coiled heat exchanger, or a finned tube heat exchanger.

This method ensures that more water is in contact with the outer wall of the heat exchanger, vastly improving the heat transfer between the primary and secondary water.

Image 5a Example of high performance storage cylinder

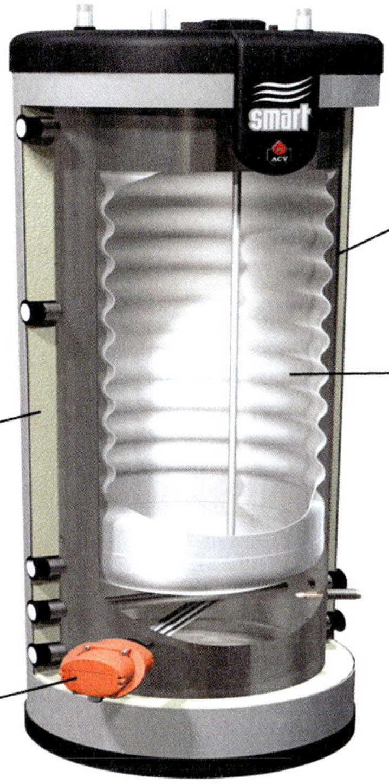

Image 5b Example of tank in tank hot water cylinder

An alternative high efficiency storage cylinder is the tank in tank system. In this the whole of the outside domestic hot water store is in contact with the primary hot water flow giving a fast heat recovery. Also because the store is flexible and expands and contracts, it helps to prevent limescale formation.

All hot water storage cylinders should be capable of losing as little heat as possible and recovering to 60°C quickly when water is used and the larger the heat exchanger surface the quicker heat can be transferred to the stored water.

All manufacturers produce cylinders in a variety of sizes to suit individual installations.

Water jacketed tube heater or thermal store

Hot water thermal store units look very similar to combination storage units. Thermal stores are supplied directly from the incoming cold water main.

Thermal stores are suitable for new and existing installations. They are available in various sizes and capacities, manufactured from copper or stainless steel, and can be of either direct or indirect type.

Thermal stores provide heated water on a directly opposite basis to that of a standard hot water installation.

Direct type

Direct thermal stores are fitted with a single integral coil. Mains cold water enters the coil via the lower tapping. The primary water contained within the vessel is heated directly by the boiler, and this water heats the water in the coil as it passes through it.

Hot water is distributed to the draw off points from the top tapping of the coil, passing through the blending valve, which ensures that the distribution temperature doesn't exceed 55°C.

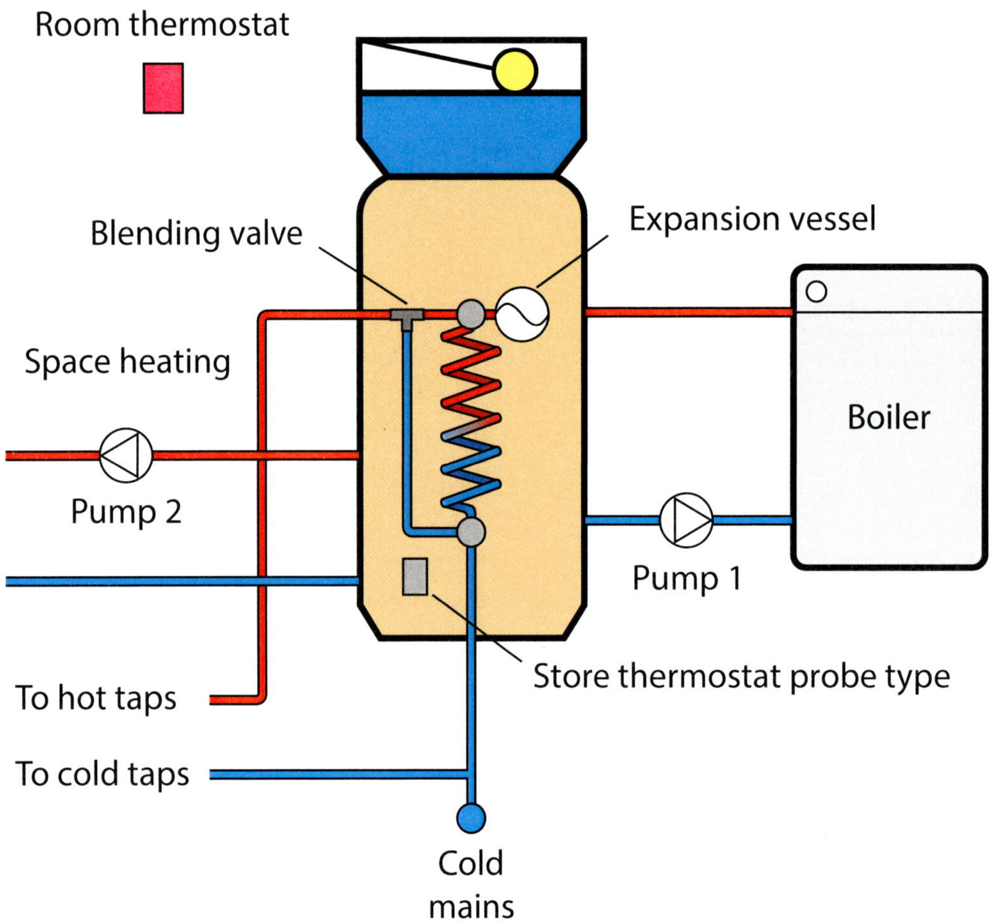

Image 6 **Direct thermal store**

Direct thermal stores are designed for open vented primary systems and are not suitable for primary sealed systems.

Indirect type

Indirect thermal stores are fitted with two internal coils; the water contained within the vessel is heated indirectly by the boiler via the lower multi-coil heat exchanger. The stored water within the unit is static, and doesn't mix with the primary water.

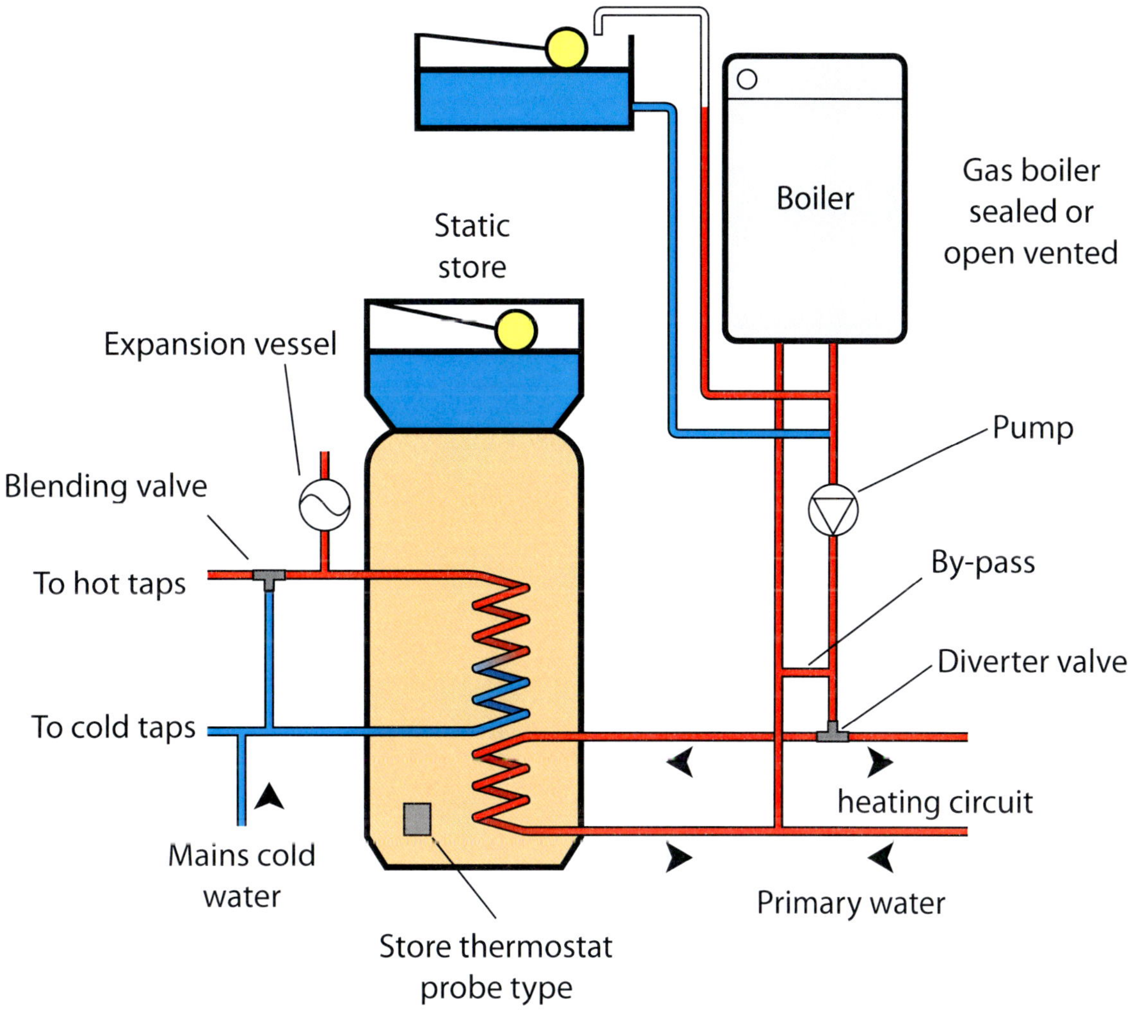

Image 7 Indirect thermal store

The upper coil provides hot water to the draw off outlets, working on the same principle as a direct system, with the addition of an expansion vessel fitted before the blending valve.

Why might a thermal store unit be installed, and what are its advantages?

Indirect thermal stores are suitable for both open vented primary systems and sealed systems, and because the static water store is open vented, they are not required to be installed at the highest point of a system, so this gives greater flexibility when siting the units.

The storage capacity for an average three bedroom domestic dwelling with one bathroom and thermostatic shower would be a 120 or 140 litres.

Both direct and indirect systems (supplementary hot water) can be heated by an immersion heater, this is not shown on the images.

Other forms of hot water supply

Because of the requirements of G3, this course now encompasses all hot water systems including water heaters with a capacity of 15 litres or less. G3 states:

'Water heaters that have a capacity of 15 litres or less that have appropriate safety devices for temperature and pressure will generally satisfy the requirement set out in G3 (3)'.

Here's an overview of other forms of hot water heating.

Unvented hot water systems under 15 litres

These can be categorised as:

- Unvented single installations
- Unvented multiple installations
- Open outlet installations

Unvented single installations

Gas fired

Gas fired sink water heaters provide instant hot water to a single point at a selected temperature.

These are no longer available for the natural gas market but may be encountered in LPG forms in caravans or residential park homes.

The following diagrams are illustrations and not technical installation instructions. Always refer to the manufacturer's instructions for accurate guidance.

Electric

The image below shows a typical installation for an under sink installation.

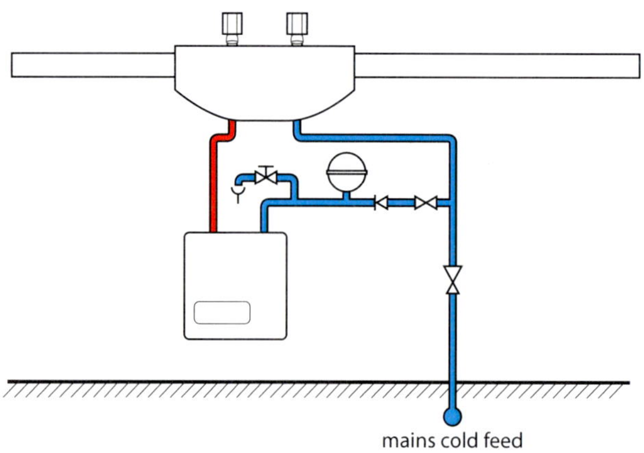

Image 8 Under sink water heater

Unvented multiple installations

The water heater can also be either gas or electrically heated.

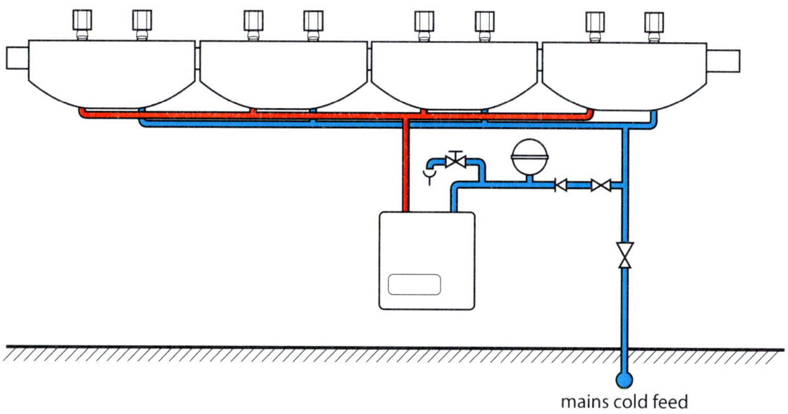

Image 9 **Electric multiple installation**

This type of heater was designed to provide water to more than one point, typically kitchen sink, bath, and wash hand basin. It was ideal for small dwellings.

They are less popular today as combination boilers have replaced them in most dwellings but they may be found in washrooms etc.

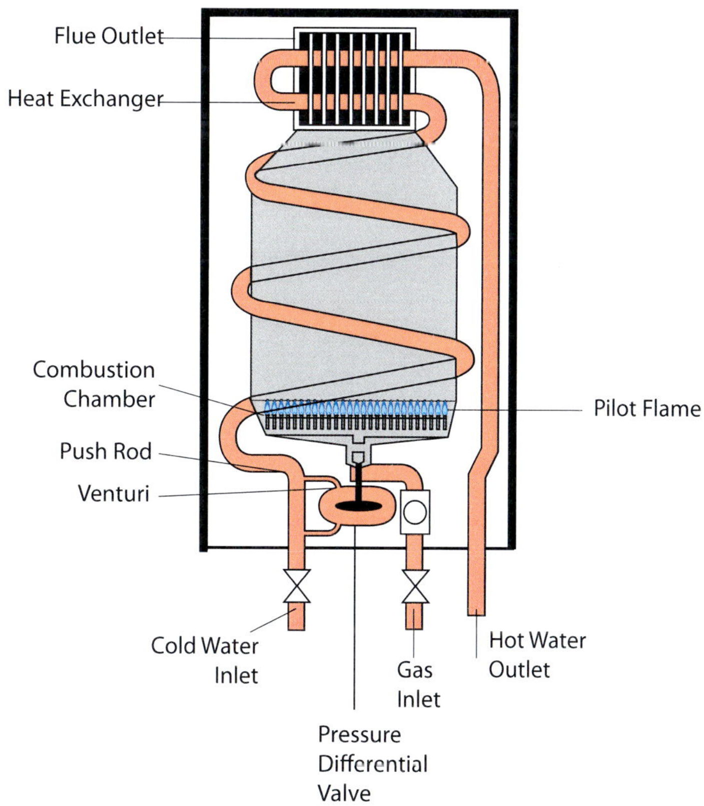

Image 10 **Typical gas multipoint water heater**

Open outlet installation

This type of heater is designed to provide water at a single point, normally over a bath or kitchen sink, and is heated by electric. Hot water is usually provided through a swivel spout outlet. It is an ideal installation for an office, kitchen or single WC.

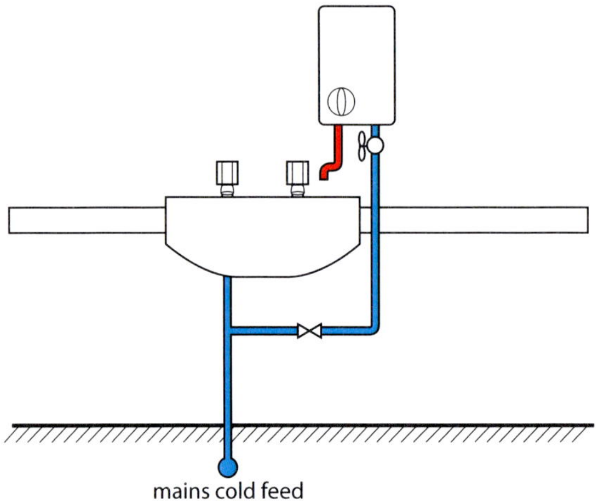

Image 11 **Open outlet heater**

Unvented implications

So what happens to stored water when its temperature exceeds 100°C?

As water pressure increases so does the boiling point of water.

As you can see from the graph below, the boiling point of water at 3.5 bar, (which is the operating pressure of many unvented systems), is 147.7°C.

The boiling point of water at elevated pressures

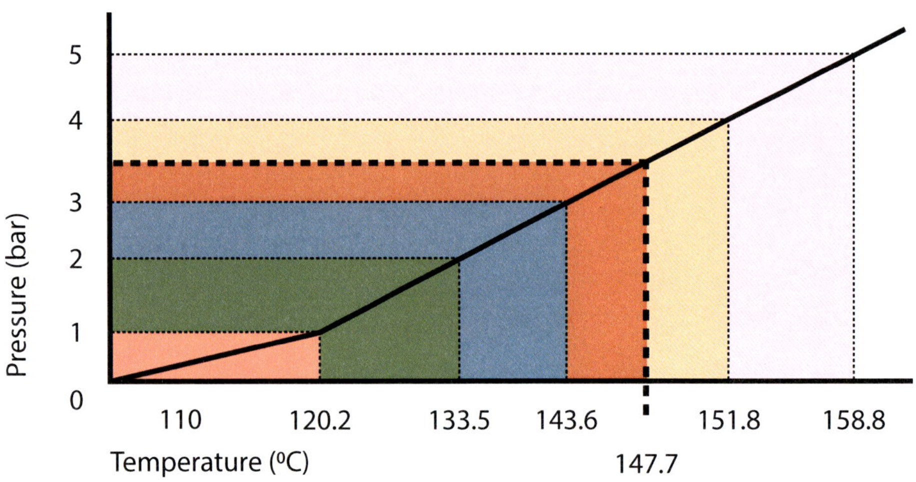

Water stored at above atmospheric pressure and at a greater temperature than 100°C, if suddenly exposed to atmospheric pressure (an open tap or fractured pipe or cylinder), the excess energy stored in the water would be released in the form of flash steam.

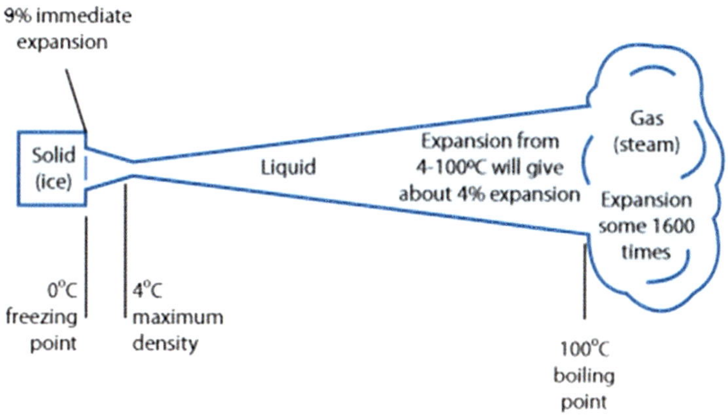

When water changes its state from liquid to gas in the form of flash steam its volume increases by up to 1600 times.

So, let's apply this information to our original calculation, roughly speaking

200 litres x 1600 = 320,000 litres of vapour!

There is obviously no way in which that type of volume can be taken up in the system and there is only one result; **explosion!**

Image 12 **Exploded cylinder**

To the left, the outcomes of an incorrectly installed unvented hot water storage cylinder

So, to recap, the purpose of the open vent pipe is to relieve any pressure build up in the system:

- Should the system overheat
- Should the normal route for expansion, the feed pipe, become blocked or closed
- Should the open vent become blocked
- Or a combination of all three

Note

The sizing of expansion vessels or expansion chambers for unvented systems involves complex calculation procedures and is usually undertaken by the manufacturer on designing the unvented system, the manufacturer should therefore always be consulted as to size of vessel or chamber. The calculations on the previous pages are included for illustrative purposes only.

Unvented systems

Now we've covered traditional vented systems, let's take a look at an unvented system.

Image 13 Unvented System Components

This system above utilises an expansion vessel.

What do you think are the basic differences between the vented and unvented systems?

One of the differences is evident in the title, but in summary, unvented systems differ from vented systems in that:

- Unvented systems are fed from the mains supply, so the rate of discharge (water flow rate) and water pressure at the tap are largely determined by the incoming supply mains pressure and water flow rate; with potentially higher flow rates and pressures available at the tap.
- In the unvented hot water storage system the expansion of hot water due to the heating effect is taken up by an expansion vessel, or an air gap contained in the top of the cylinder.

The unvented system equivalent of the open vent pipe is the temperature relief valve which is designed to discharge hot water safely to atmosphere should the water temperature in the system increase above a pre-determined level.

Remember:

In the vented system, expansion is catered for via the 'non-mechanical' use of the cold feed/storage cistern and the open vent pipe arrangement for safety purposes.

In the unvented system, mechanical control devices are used and these can be more prone to failure than the non-mechanical equivalent in the vented system, therefore a series of additional control devices are necessary. It's important therefore to ensure a tight control on who may install or maintain unvented systems.

Advantages and disadvantages of unvented hot water storage systems over other forms of hot water storage and supply

Advantages

The following advantages assume that mains pressure and flow rate are satisfactory

Features	Benefits
High water pressure	High flow rates to ALL terminal fittings. Shower delivery rates improved, no need for pumps. Baths and sinks fill more quickly.
Balanced hot and cold supplies	Improved performance at mixer taps and showers. Improved customer safety. Potential for use of 'luxury' quality terminal fittings.
No cistern in roof space	No risk of freezing in winter. Less risk of contamination. Potential for alternative use of roof space. Quieter system, no noisy float valves. Safer to work on, no access to roof space required.
Easier to install	Although purchase cost greater, installation costs less.
Flexibility of cylinder location (influenced by discharge pipework)	Improved pipe runs. Less chance of 'dead legs'. May free up 'premium space'.

Disadvantages

Features	Results
Low incoming mains water pressure/flow rate	Insufficient pressure and flow rate for system to work correctly
Where future mains pressure reductions may be likely	Insufficient pressure and flow rate for system to work correctly
No water storage back up	No water when mains fails
Installation requires visible discharge pipe	Can be difficult to site to maintain required flow rate
Extra pipework and controls	Takes up more space, may be all of the cylinder cupboard
Purchase price of unvented cylinder	Usually costs more to purchase
On going costs	Service and maintenance, training requirement

So now we've covered the basics let's get into a little more detail.

The legislation covering the installation and maintenance of hot water storage systems

There are two specific areas of legislation covering the installation and maintenance of unvented hot water storage systems in England & Wales:

- Building Regulations – Approved Document G section 2 Water efficiency
- Building Regulations – Approved Document G section 3 Hot water supply and systems
- Water Supply (Water Fittings) Regulations 1999
- Building Regulations – Approved Document L1A Conservation of fuel and power in new dwellings
- Building Regulations – Approved Document L1B Conservation of fuel and power in existing dwellings
- Building Regulations – Approved Document P1 Design and installation of electrical installations
- Domestic Building Services Compliance Guide

It should be noted that there is some overlap between both sets of legislative requirements.

Additionally, for unvented systems, the cylinder must be approved by a certification body having accreditation from the United Kingdom Accreditation Service (UKAS) or an equivalent European body. In practice this usually means BBA, KIWA or WRc approval.

Our aim in this training course is to provide a 'user friendly' interpretation of those requirements so that you are able to comply with the legislation.

The Building Regulation G3 deals primarily with the installation and safety related issues of hot water storage systems whereas the Water Regulations primarily deal with system functional issues including waste, misuse, contamination and undue consumption of the water supply. Building Regulation G2 also deals with water efficiency.

Building Regulations Part L1A and L1B deal with energy efficiency and prevention of wastage. The Domestic Building Services Compliance Guide is basically complementary guidance on how to comply with Part L.

Building Regulations Part P deals with the electrical safety issues surrounding supplies and connections to hot water storage systems.

Building Regulation requirements

Limits on application

The Building Regulation requirement G2 only applies when a dwelling is:

- Erected; or
- Formed by a material change of use of a building

The Building Regulation requirement G3(4) only applies when a dwelling is:

- Erected; or
- Formed by a material change of use of a building

The Building Regulation requirement G3(3) does not apply to:

- A system which heats or stores water for the purposes only of an industrial process.

Regulation G2: Water efficiency including regulation 17K and 20E

Regulations 17K and 20E are part of the general building regulations, which can be found at:

http://www.planningportal.gov.uk/england/professionals/buildingregs/

The bulk of the requirements within the Building Regulations are grouped according to performance topic (e.g. sanitation is in 'Group G'). All the grouped requirements are listed within schedule 1 of the Building Regulations.

These same grouped requirements are then repeated in Documents like Approved Document G, where additional guidance is provided.

Requirements

G2: The fittings and appliances used must help prevent unnecessary water consumption. This requirement applies to new build and building conversions (e.g. barn conversion to dwelling – buildings which have undergone a material change of use).

17K The 'Water Efficiency Calculator for New Dwellings' must be used to estimate consumption. A limit of 125 litres per person per day is set.

20E A notice giving the result of this calculation must be supplied to the Local Authority within 5 days of the work being completed.

Guidance

Manufacturers of sanitary appliances and white goods need to declare their products' water consumption figures. The products chosen to achieve the overall maximum limit must be fitted and instructions on use and maintenance given to the building owner.

5 litres/person/day is factored in for outdoor use, leaving 120 litres for internal consumption. Completion certificates should only be issued by Building Control Bodies once they have received an acceptable water calculation tool result.

The Water Efficiency Calculator for new dwellings

The Water Energy Efficiency Calculator is structured in two main sections:

- Water Calculator Methodology
- Calculation Tables

Water Calculation Methodology

The document sets out the Water Calculation Methodology for assessing the whole house water efficiency of new dwellings. The calculation method is to be used to assess compliance against the water performance targets in Building Regulations Part G as set out below.

It is not a design tool for water supply and drainage systems. It is also not capable of calculating the actual water consumption of a new dwelling. Behaviour and changing behaviour can also have an effect on the amount of wholesome water used throughout a home.

Performance target	Maximum consumption (litres/person/day)
Part G/Regulation 17.K compliance	125

The calculation method requires the use of water consumption figures provided from manufacturer's product details. Before the assessment can be carried out, figures will need to be collected from manufacturer's product information to determine the consumption of each terminal fitting.

G2 Specifies a maximum daily usage of 125 litres per person which must be designed at the planning stage for all new builds.

Swimming pools, Jacuzzis and hot tubs are specifically excluded. Compliance can be proved to local building control by submitting a copy of the design calculation sheet. For washing machines etc, the manufacturer's default value should be used. Super efficient terminal fittings can be offset against water guzzlers provided the 125 litres per day is not exceeded.

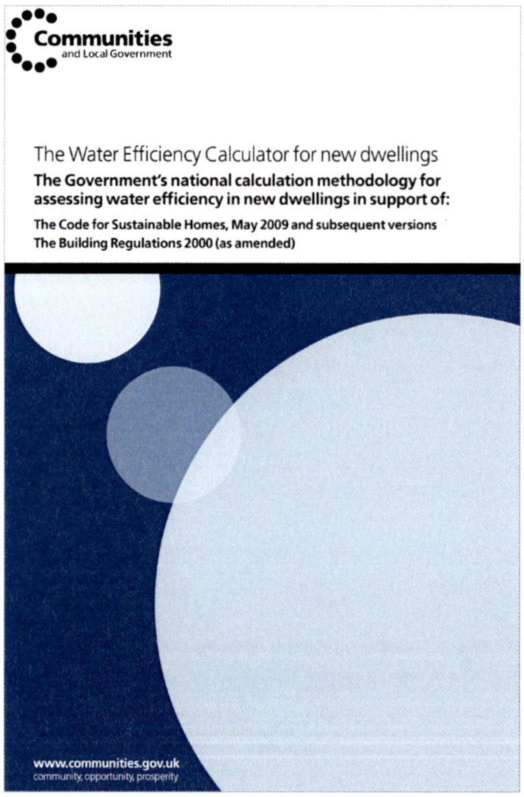

Water calculation tables

The tables are essentially a number of forms to complete much like an annual tax return!

Capacity/flow rate

In the case of items such as toilets and baths etc a figure for the average volume of water consumed per use is entered. For items such as taps it is the flow rate in litres per minute. These figures need to be supplied by the manufacturers.

Use factor

This is a figure that gives a standard number of times that the device is used per person per day. Using this table the user calculates the expected volume of water consumed per person per day, which must not exceed 125, including a set figure of 5 litres for external use.

The Water Efficiency Calculator can be found at http://www.planningportal.gov.uk/uploads/br/water_efficiency_calculator.pdf

In situations where the equipment installed is not as specified in the design specification, the daily per head use calculations will have to be re-calculated and submitted for approval.

Extract from calculator table showing type of information required.

Table 1: The water calculator for new dwellings

Installation type	Unit of measure	Capacity/ flow rate (1)	Use factor (2)	Fixed use (litres/ person/ day) (3)	Litres/ person/day = [(1) x (2)] + (3) (4)
WC (single flush)	Flush volume (litres)		4.42	0.00	
WC (dual flush)	Full flush volume (litres)		1.46	0.00	
	Part flush volume (litres)		2.96	0.00	
WCs (multiple fittings)	Average effective flushing volume (litres)		4.42	0.00	
Taps (excluding kitchen/ utility room taps)	Flow rate (litres/minute)		1.58	1.58	
Bath (where shower also present)	Capacity to overflow (litres)		0.11	0.00	
Shower (where bath also present)	Flow rate (litres/minute)		4.37	0.00	
Bath only	Capacity to overflow (litres)		0.50	0.00	
Shower only	Flow rate (litres/minute)		5.60	0.00	
Kitchen/utility room sink taps	Flow rate (litres/minute)		0.44	10.36	
Washing machine	Litres/kg dry load		2.1	0.00	
Dishwasher	Litres/place setting		3.6	0.00	

Requirements of Regulation G3

Regulation G3: Design and installation of directly or indirectly heated hot water storage systems

General

- The Regulation G3 states the hot water storage systems should be designed and installed in accordance with BS 6700: 2006 Specification for design, installation, testing and maintenance of services supplying water for domestic use within buildings and their cartilages or BS EN 12897: 2006 Water supply. Specification for indirectly heated unvented (closed) storage water heaters.
- Reference is also made to the requirements for hot water storage vessels in terms of their conformity to British Standards.

Vented hot water storage systems

3.12 Vented hot water storage systems should incorporate a vent pipe of an adequate size, but not less than 19mm internal diameter, connecting the top of the hot water storage vessel to point open to the atmosphere above and over the level of the water in the cold water storage cistern.

3.13 In addition to the vent pipe referred to in 3.12 and any thermostat provided to control the temperature of the stored water to a desired temperature, vented hot water storage systems should incorporate either:

a. i) For all direct heat sources, a non-self resetting energy cut-out to disconnect the supply of heat to the storage vessel in the event of the storage system overheating; and

ii) For all indirect heat sources, an overheat cut-out to disconnect the supply of heat to the storage vessel in the event of the stored water overheating so that the temperature of the stored water does not exceed 100°C; or

b An appropriate safety device, for example, a temperature relief valve or a combined temperature and pressure relief valve to safely discharge the water in the event of significant overheating.

3.14 Vent pipes should discharge over a cold water storage cistern conforming to BS 417-2: 1987 or BS 4213: 2004

3.15 The cold water storage cistern into which the vent pipe discharges should be supported on a flat, level, rigid platform which is capable of safely withstanding the weight of the cistern when filled with water to the rim and fully supporting the bottom of the cistern over the whole of its area. The platform should extend a minimum of 150mm in all directions beyond the edge of the maximum dimensions of the cistern.

Note: Where an existing metal cistern is replaced, or a plastic cistern is replaced by one with larger dimensions, the existing support should be upgraded, as necessary, with one in accordance with paragraph 3.15

3.16 The cistern should be accessible for maintenance, cleaning and replacement

Note: Although Building Regulation Approved Document G still refers to BS 6700: 2006 this has been superseded by BS EN 806 Parts 1 to 6 and BS 8558: 2011.

Example of good practice of cistern support, the platform should be extended a minimum of 150mm in all directions beyond the edge of the maximum dimensions of the cistern.

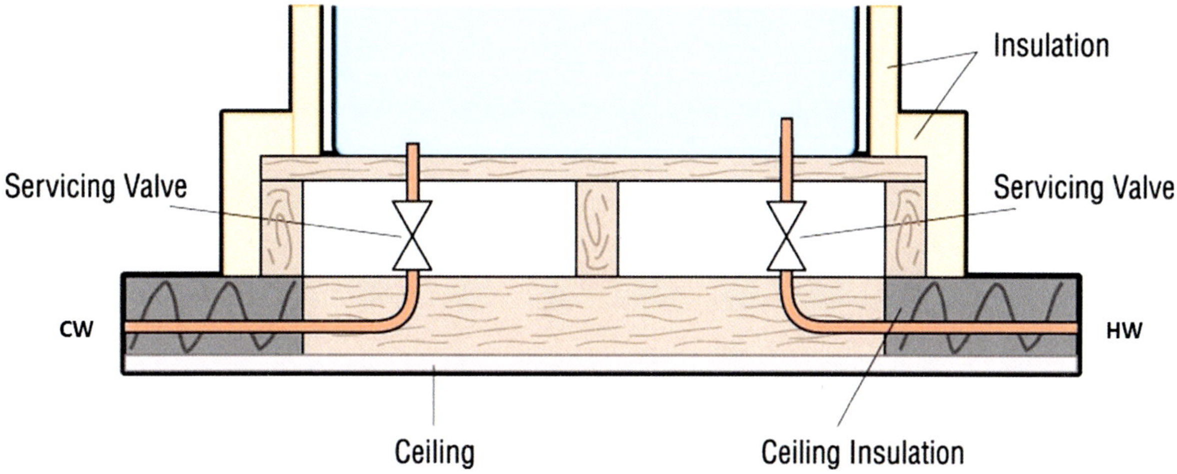

Image 14 Cistern support

Hopefully, adhering to these regulations will prevent disasters like the following extract from a press report in 2002:

Mother of two dies after scalding

A mother-of-two from Cornwall died after scalding water poured out of a water tank through a ceiling and on to her bed as she slept, an inquest has heard. She was admitted to hospital with 45% burns after the accident at her housing-authority owned home and died from her injuries nine days later.

Tests carried out by the Health and Safety Executive (HSE) later showed that a fault in the property's immersion heater had allowed the cold water tank to fill with scalding water. The house was gas centrally heated, but the electric immersion heater was used for the hot water.

The HSE inspector told the inquest he had carried out tests on the cold water tank from the loft and the immersion heater's hot water tank from the women's bedroom. He said a faulty thermostat meant the water was able to reach boiling point instead of stabilising at 60°C.

The hot water then flowed into the cold water tank in the loft until eventually that too was filled with boiling water.

The inspector said: "I understand that the tank wasn't fully supported. In my view this would explain why this tank slumped and released its water", he added "it was not clear why the thermostat had failed".

Unvented Hot Water Storage Systems and G3

Regulation G3: Design and installation of directly or indirectly heated hot water storage systems

What the regulations say

Regulation G3: Unvented hot water storage systems – all systems

3.17 To minimise the danger from excessive pressure, unvented hot water storage systems should incorporate a minimum of two independent safety devices.

These shall be in addition to any thermostat provided to control the temperature of the stored water. The selection of safety devices should take account of the physical location of the devices, and the design, configuration, location of components and performance characteristics of the system to which they are attached.

3.18 An acceptable approach might consist of:

a. A non-self resetting energy cut-out to disconnect the supply of heat to the storage vessel in the event of the storage system overheating; and

b. A temperature relief valve or a combined temperature and pressure relief valve to safely discharge the water in the event of serious overheating.

Alternative approaches to this are acceptable provided that they provide an equivalent degree of safety.

3.19 Water heaters with a capacity of 15 litres or less that have the appropriate safety devices for temperature and pressure will generally satisfy the requirement set out in G3(3)

Unvented hot water storage systems – systems up to 500 litres capacity and 45kW power input

3.20 Paragraphs 3.21 to 3.24 are in addition to the provisions of 3.17 above.

3.21 If an indirect supply of heat to an unvented hot water storage system incorporates a boiler, the energy cut-out may be on the boiler.

3.22 Any unvented hot water storage system up to 500 litres and less than 45kW should be in the form of a proprietary hot water storage system unit or package.

The package and components should be appropriate to the circumstances in which they are used and should satisfy any appropriate standard that will ensure the requirements of regulation G3(2) and G3(3) will be met *(e.g. BS 12897: 2006 Water Supply. Specification for indirectly heated unvented (closed) hot water storage systems or BS 6700: 2006 + A1 2009 Design, installation, testing and maintenance of services supplying water for domestic use within buildings and their curtilages).*

3.23 Any unvented hot water storage system unit or package should be indelibly marked with the following information:

a. The manufacturer's name and contact details;
b. A model reference;
c. The rated storage capacity of the storage water heater;
d. The operating pressure of the system and the operating pressure of the expansion valve;
e. Relevant operating data on each of the safety devices fitted; and
f. The maximum primary circuit pressure and flow temperature of indirect hot water storage system units or packages

The unvented hot water system design

Regulation G3: *Continued*

3.24 In addition, the following warning should be indelibly marked on the hot water storage system unit or package so that it is visible after installation:

WARNING TO USER

a. Do not remove or adjust any component part of this unvented water heated; contact the installer.

b. If this unvented water heater develops a fault, such as flow of hot water from the discharge pipe, switch the heater off and contact the installer.

WARNING TO THE INSTALLER

a. This installation is subject to Building Regulations.

b. Use only appropriate components for installation or maintenance.

Installed by:

Name: ....

Address: ....

Tel. No. ....

Completion date: ....

Unvented hot water storage systems – systems over 500 litres capacity or over 45kW power input

3.25 Paragraph 3.26 and 3.27 are in addition to the provisions of 3.17.

3.26 Systems over 500 litres capacity will generally be bespoke designs for specific projects and as such are inappropriate for approval by a third party accredited product conformity certification scheme. Where this is the case, the unvented hot water system should be designed to the safety requirements in 3.17 by an appropriately qualified engineer.

3.27 Any unvented hot water storage system having a power input of more than 45kW, but with a capacity of 500 litres or less should be in the form of a proprietary hot water storage unit or package.

The package and components should be appropriate to the circumstances in which they are used and should satisfy an appropriate standard that will ensure the requirement of regulation G3(2) and G3(3) will be met *(e.g. BS 12897: 2006 Water Supply. Specification for indirectly heated unvented (closed) hot water storage systems or BS 6700: 2006 + A1 2009 Design, installation, testing and maintenance of services supplying water for domestic use within buildings and their curtilages).*

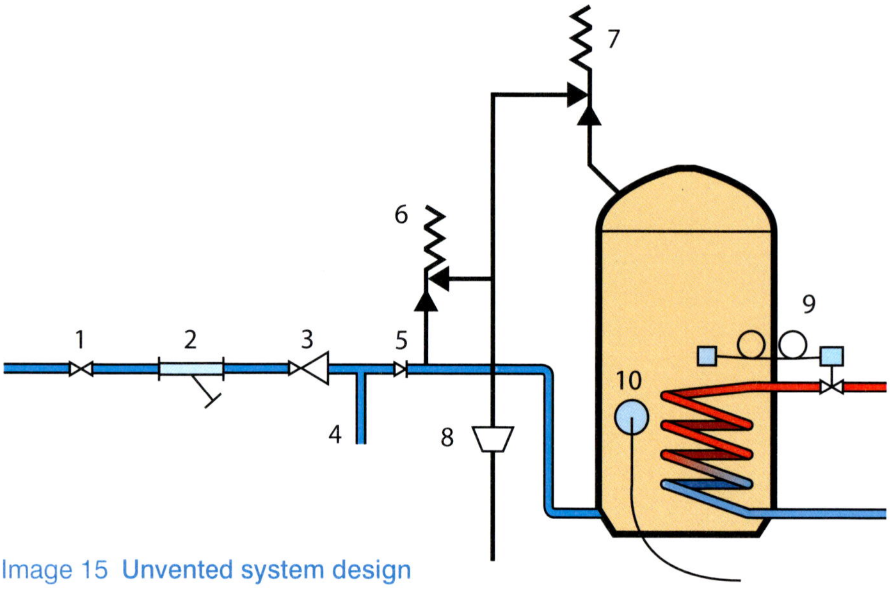

Image 15 **Unvented system design**

Image 15 shows how the early design of system controls were laid out. Modern systems retain the same order but a number of the controls are assembled into a combination valve.

1. Isolation valve on cold supply to cylinder
2. Strainer
3. Pressure Reducing Valve (PRV)
4. Balanced cold supply take off point
5. Single check or non-return valve
6. Expansion (pressure) relief valve
7. Temperature and pressure relief valve
8. Tundish
9. Combined cylinder thermostat and Energy Cut Out (ECO) controlling 2 port motorised valve
10. Immersion heater with non-self resetting Energy Cut Out

The following diagram shows a typical modern control layout.

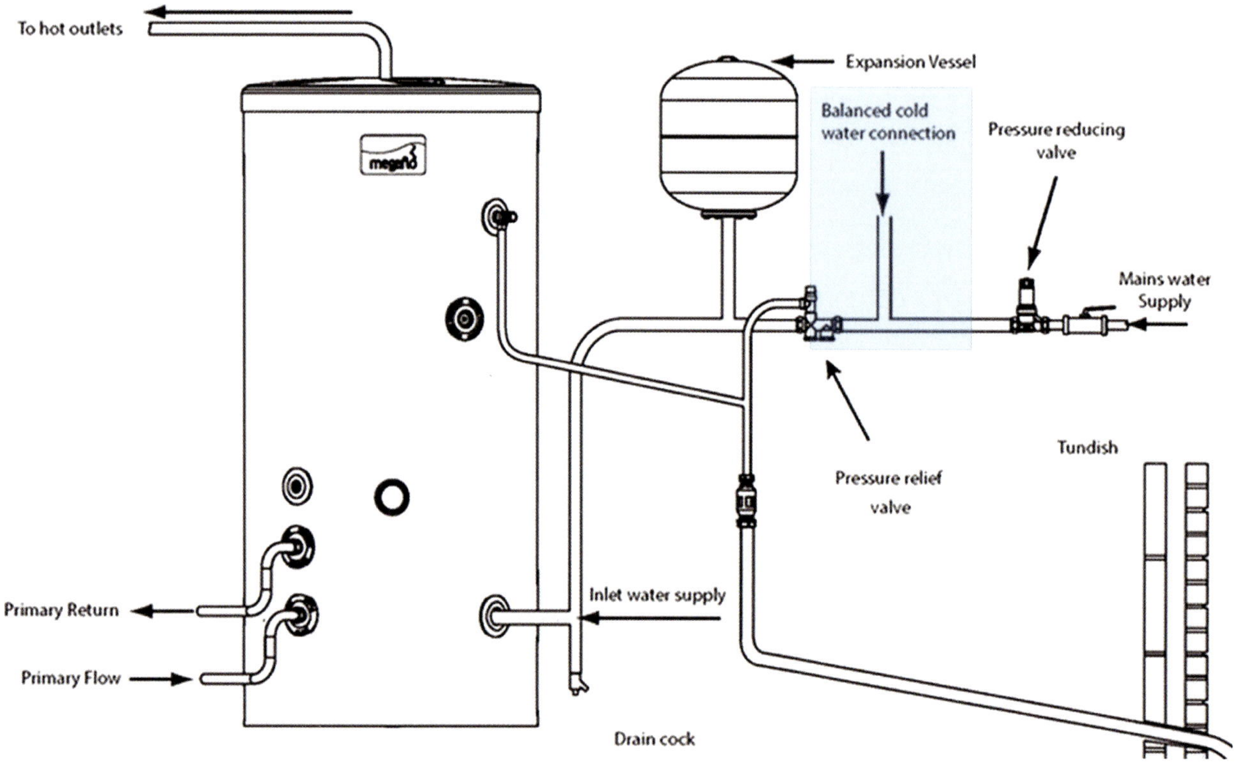

Image 16 **Modern unvented system control layout**

Let's take a trip round the components!

Safety devices

Control thermostat	Controls the stored water temperature to between 60 – 65°C. The immersion heater will have one built in and a separate one will be supplied for indirect cylinders. Both will have a built in Energy Cut Out (ECO).
Energy Cut-Out (ECO) (High limit thermostat)	This device is designed to isolate the electrical supply to the immersion heater via the built in non-self resetting cut out. For indirect systems it turns off the heat source to an indirect cylinder by closing a 2 port motorised valve using a non-self resetting cut out. It has a maximum setting of 90°C but usually operates at around 85°C.
Temperature & Pressure relief valve	This valve is often referred to simply as the temperature relief valve but it will also operate as a pressure relief valve in the event of the failure of the pressure relief valve. It is designed to open to discharge water when the temperature of the stored water is 90 ± 5°C .

The above safety devices grouped together in an unvented system are often referred to as a **three-tier level of safety protection.**

Functional devices

Isolating valve	This is required to turn off the water supply in order for maintenance to be carried out without undue wastage of water. At least one manufacturer includes this in the combination valve.
Combination valve	Modern unvented hot water storage systems usually have a combination valve which contains the following: • Line strainer • Pressure reducing valve • Pressure relief valve • Non-return or check valve Older systems often have separate controls installed in the order listed.
Line strainer	This is required to prevent particles entering the pipework and controls which may prevent them operating correctly.

Pressure reducing valve	The pressure of the incoming service can vary and this valve is used to set and maintain a constant operating pressure within the system.
Pressure relief valve	Designed discharge water from the system at a pre-determined pressure to prevent high pressure damage to the operational controls. It will discharge in the event of failure of the pressure reducing valve or the expansion vessel or bubble.
Single non-return or check valve	This valve prevents stored hot water or potentially contaminated balanced cold water being siphoned or forced back into the supply and being drawn off as drinking water.

Separate controls	
Expansion device (External vessel or internal air bubble)	Required to accommodate the expansion of the stored water when heated. Some systems will have a separate external expansion vessel, some systems utilise an air gap trapped at the top of the cylinder.
Tundish	Required as a visible means for the customer to identify when either the temperature or expansion relief valve is discharging and to provide a relief opening in the event that the discharge pipe becomes blocked.

Now that we are familiar with the component parts let's look at them as they apply to the various types of unvented system in the context of the two items of legislation; building and water.

So what are the types of unvented system?

Systems fall into two categories:

- Storage (over 15 litres capacity)
- Independent water heaters (sometimes called point of use or multipoint heaters under 15 litres capacity)

Point of use or multipoint water heaters

Including combination boilers and thermal storage units with a storage capacity of less than 15 litres will generally satisfy the requirement set out in G3(3) if they have appropriate safety devices for temperature and pressure.

Remember: Combination boilers of the high flow type may well include more than 15 litres of stored water and will therefore be subject to the following requirements.

Storage systems

Storage systems fall into two categories:

- **Directly heated systems** – where the water is directly heated by e.g. a gas burner (as shown below) or an immersion heater

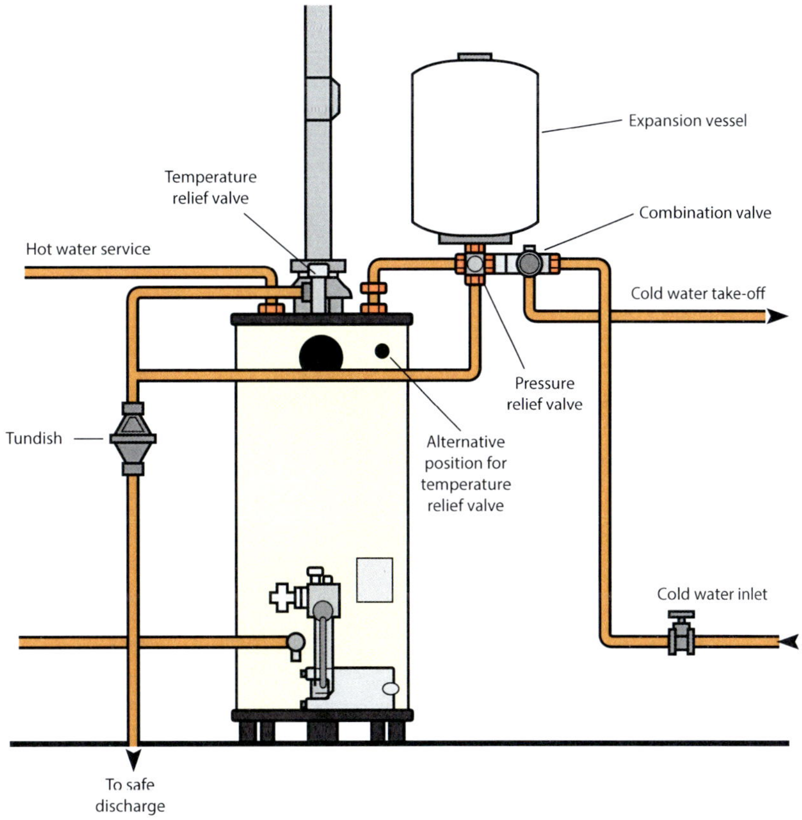

Image 17 Directly heated system

- **Indirectly heated systems** – where the water is indirectly heated through a coil inserted inside the cylinder

Image 18 **Indirectly heated system**

Building regulation requirements – summary

We've covered the regulations in quite a bit of detail, but here's a quick reminder.

Remember an unvented hot water storage system is a system that has a storage cylinder which does not incorporate a vent pipe to atmosphere.

The key requirements are:

- The system shall be installed by a person competent to do so;
- There shall be precautions to:
  - Prevent the temperature of stored water at any time exceeding 100°C, and
  - Ensure that hot water discharged from safety devices is safely conveyed to where it is visible but will not cause danger to persons in or about the building.

Preparing to undertake unvented system installation

The installation of an unvented hot water storage system requires notice to be given to the local authority building control department by means of a Building Control Notice.

This is not required if the installation has been carried out by an individual/business who has registered with a competent person scheme for the purposes of self certifying unvented hot water storage system installations.

There are a number of these schemes which can be accessed via the BPEC website http://www.bpec.org.uk.

Now the detail.

Package or Unit?

The Regulations state that an unvented hot water storage system should be in the form of a unit or package.

A '**unit**' (shown right) means an unvented hot water storage system which has all the safety devices and functional (operational) devices factory fitted by the manufacturer.

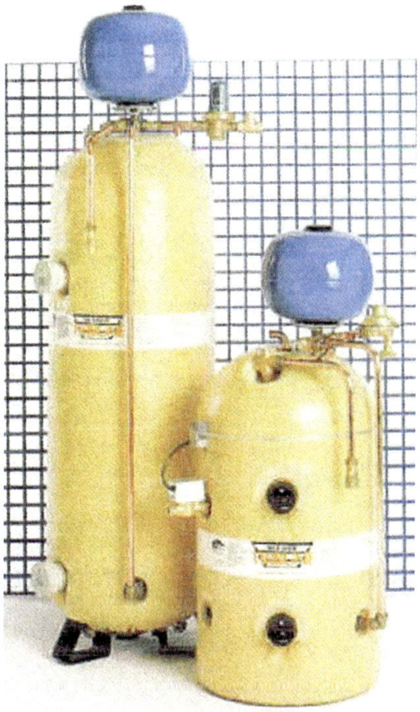

Image 19 **Unvented unit**

To include:

- Line strainer
- Pressure reducing or limiting valve
- Single check valve
- Tundish
- Expansion vessel
- Pressure relief valve
- Temperature relief valve
- High limit thermostat and cylinder primary motorised valve (indirect unit)
- Immersion heater and built in energy cut out

(All above factory fitted)

Although Approved Document G3 still refers to 'units', very few are available on the market these days. The example shown above was based on a copper cylinder and came ready assembled with only the cold supply via an isolation valve and the discharge pipework being supplied on-site. The one on the right was based on a stainless steel cylinder. The mains isolation valve may or may not have been supplied, but one must be fitted in all cases.

Image 20 **Stainless steel unit**

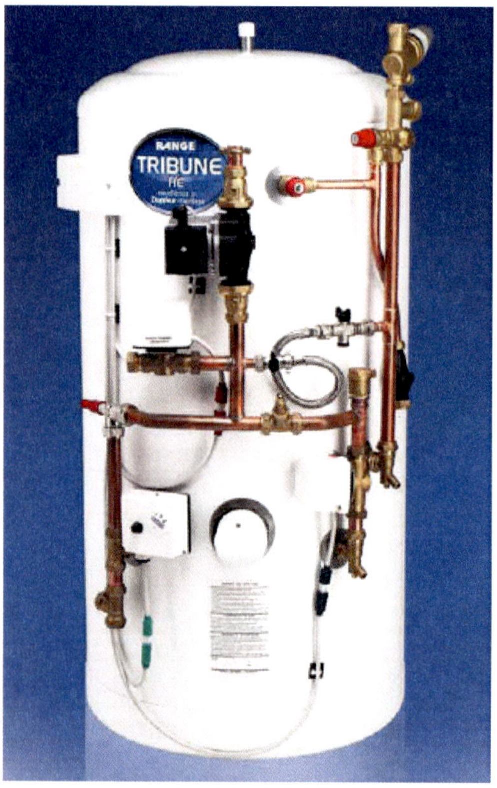

Image 21 System fit unit

System fits

Currently the bulk of 'units' come in the form of 'system fits'. This is an unvented cylinder with all the safety controls already installed by the manufacturer along with all the appropriate central heating system pump and controls.

All that is needed is to connect the primary and central heating flows and returns, cold supply and hot draw off and the discharge pipework; this can save a huge amount of time when installing a full system, hence the name.

A '**package**' (shown right) means an unvented hot water storage system having the safety devices factory fitted together with a kit containing other applicable devices, supplied by the package manufacturer, to be fitted by the installer.

Items factory fitted

- Temperature relief valve
- Immersion heater, thermostat and built in energy cut out
- Thermostat and energy cut out for indirect heat source (may be supplied separately for installer to fit)

Items supplied by the manufacturer, to be fitted by the installer

- Line strainer
- Cylinder primary motorised valve (indirect unit)
- Pressure reducing or limiting valve
- Single check valve
- Tundish
- Expansion vessel
- Pressure relief valve

Image 22 Unvented package

Modern packages comprise a cylinder with a factory fitted temperature relief valve and thermostat/energy cut-off; the tundish, non-return valve and pressure reducing valve or combination valve are supplied separately.

There may be instances where thermostats are not fitted to the cylinder but supplied by the manufacturer for the installer to insert into a built in phial pocket.

If the package is not a 'bubbletop', one with an internal expansion bubble within the cylinder, then an expansion vessel will also be supplied.

Currently approximately 90% of installations are 'packages' and of the remaining 10% of 'units', most are 'system fits'. Units are generally more expensive than packages.

Image 23 **Pressure relief valve**

Image 24 **Temperature relief valve**

Image 25 **Expansion vessel**

Image 26 **Tundish**

Image 27 **Tundish connected to relief valves**

Image 28 **Combination valve with internal check valve**

Approval of the Unit or Package

The Regulations state that any unvented hot water storage system up to 500 litres or 45kW should be in the form of a proprietary hot water storage system unit or package. The package and components should be appropriate to the circumstances in which they are used. Compliance can be demonstrated by independent certification schemes such as the following:

- Approved by a certification body having accreditation from the United Kingdom Accreditation Service (UKAS) or an equivalent European accreditation body as complying with:
  - The relevant requirement of regulation G3(3) or
  - An appropriate standard that will ensure the requirement of regulation G2(3) will be met (e.g. BS EN 12897: 2006 Water Supply. Specification for indirectly heated unvented (closed) hot water storage systems or BS 6700: 2006 Design, installation, testing and maintenance of services supplying water for domestic use within buildings and their curtilages);

Or

- The subject of an independent assessment that clearly demonstrates an equivalent level of verification and performance to that described above.

In the UK this generally means BBA, KIWA or WRc approval.

Components Required by the Building Regulations

Direct heating

A directly heated unit or package should have a minimum of two temperature activated safety devices operating in sequence.

- A non-self resetting thermal cut-out to BS 3955: 1986 – Specification for electrical controls for household and general purposes, or to BS 4201: 1984 – Specification for thermostats for gas burning appliances, and
- One or more temperature relief valves to BS 6283: Safety and control devices for use in hot water systems Part 2: 1991 – Specification for temperature relief valves from 1 bar to 10 bar, or Part 3: 1991 Specification for combined temperature and pressure relief valves for pressures from 1 bar to 10 bar.
- These devices are in addition to any thermostatic control which is fitted to maintain the temperature of the stored water (control thermostat).
- In both units and packages the temperature relief valve(s) should be located directly on the storage vessel such that the stored water does not exceed 100°C. In most cases the valve opens at 95°C. The valves should be sized to give a discharge rating in line with BS 6283. Each valve should discharge via a short length of metal pipe (D1) of a size not less than the nominal outlet size of the temperature relief valve either directly or by means of a manifold sized to accept the total discharge from the discharge pipes connected to it, through an air break over a tundish located vertically as near as possible to the valve(s) (see later diagram page 41).
- **From 1 April 2004 all immersion heaters whether on unvented or vented cylinders have required a built in non-self resetting energy cut out.**

Indirect heating

A cylinder with an indirect heating source, i.e. a gas or oil fired central heating boiler, will need all the controls listed for a direct system plus:

- A control thermostat to limit the temperature of the stored water usually to 65°C with a built in energy cut out. The energy cut out is normally wired to control a 2 port motorised valve which will close in the event of the energy cut out operating.
- If the heat source is a boiler with a high limit thermostat this may, if suitable, be used utilised as the energy cut out.

The example below shows an immersion heater suitable for an indirect unvented system, it includes a high limit thermostat (non self-resetting thermal cut-out device). The thermostat is usually purposely designed for the cylinder and replacements should be ordered via the manufacturer.

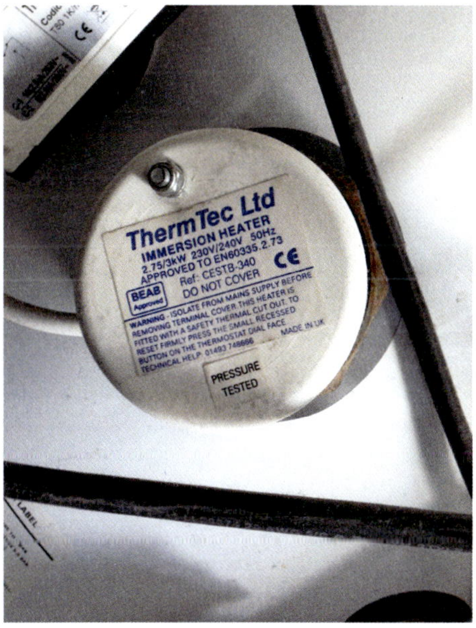

Image 29a **EN60335 Immersion heater**

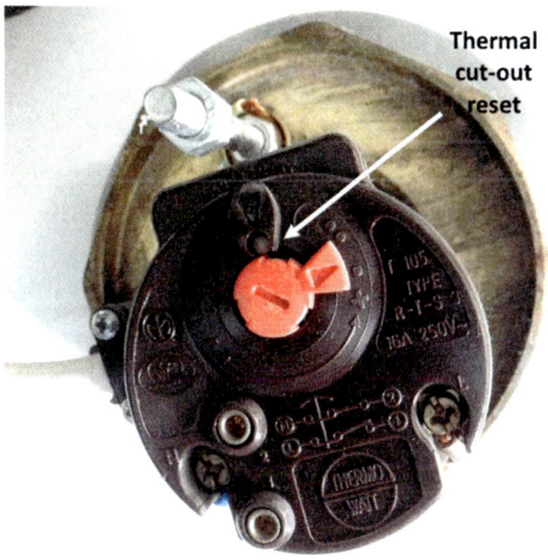

Image 29b **Thermal cut-out reset**

Competency of installers

The unit or package should be installed by a person competent in the installation of unvented hot water storage systems i.e. one holding a current operative identity card and certificate for the installation of unvented domestic hot water storage systems issued by BPEC or an equivalent body.

Image 30 BPEC's operative identity card

This card and certificate can then be used as proof of competence in order to join an appropriate Competent Persons Scheme for the self-certification of unvented hot water storage systems.

The installation of an unvented hot water storage system requires notice to be given to the local authority building control department by means of a Building Control Notice.

This is not required if the installation has been carried out by an individual/business who has registered with a competent person scheme for the purposes of self certifying unvented hot water storage system installations.

There are a number of these schemes which can be accessed via the BPEC website http://www.bpec.org.uk/services/

Your card and certificate will need to be renewed every five years except in Scotland.

If you choose not to join a Competent Persons Scheme you will need to notify your local Building Control Department of the installation before commencing work.

Discharge pipes

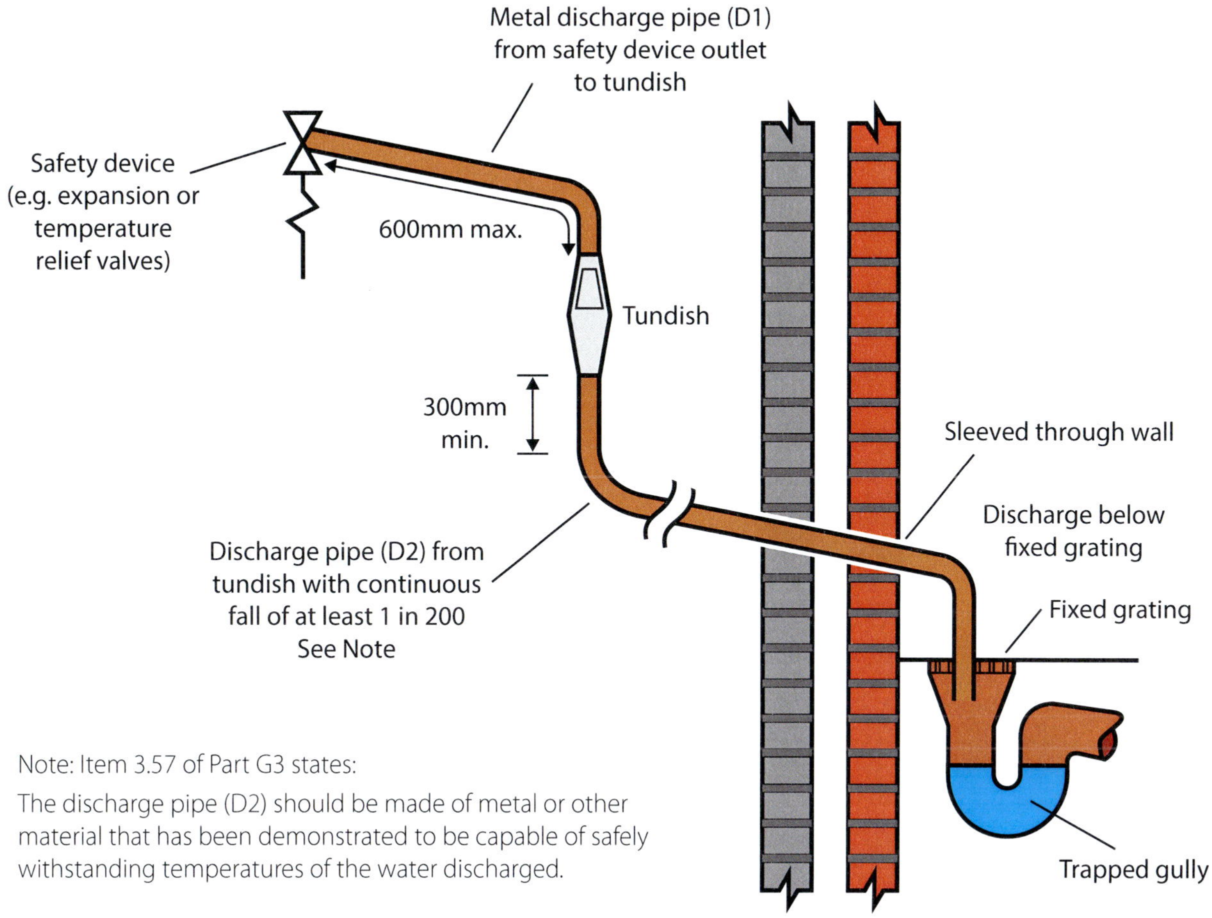

Image 31 **Discharge pipes (D1 and D2)**

The discharge pipe is divided into two sections (D1) and (D2) and is separated via a tundish. The requirements of D1 are included in this module. D2 will be covered in detail in Module 2 of this course covering design and installation issues where we deal with sizing of the discharge pipe and permissible locations for its termination.

The Regulations regarding D1 state that the safety relief valves should discharge via a short length of **metal** pipe directly or through a manifold to a tundish. If a manifold is to be used (this will be in almost every case) it must be factory fitted.

The D1 pipe should be at least the same diameter as the outlet of the safety device and be no longer than 600mm between the safety devices and the tundish.

The tundish shall be located in the same space as the cylinder, as close as possible to, and lower than the outlet of the safety relief valve or valves.

Inspection of installations

The Regulations provide guidance on the extent to which installations should be inspected by the local Building Control Department. If the installer is not a member of a competent person scheme he/she will need to notify Building Control before commencing the installation.

In this instance, it is likely that in most cases Building Control will inspect the installation. If the installer is a member of a competent person scheme he/she can notify the scheme operator upon completion of the installation. When this happens the installation may not be inspected as a matter of course but is likely be inspected as part of the quality control procedures of the scheme.

That covers the requirements of the building regulations so let's take a look at the water regulation requirements.

Water Regulation requirements

As we have indicated before the Building Regulations tend to identify the safety requirements of hot water storage systems whereas the Water Regulations deal with functional (operational) requirements primarily addressing water usage issues.

You will probably be well versed on the water regulations as they relate to vented systems, so next we will look at the key requirements of the Water Regulations in relation to unvented systems.

Every unvented storage system should be fitted with:

- A temperature control device; and
- Either a temperature relief valve or combined temperature and pressure relief valve; and
- An expansion relief valve; and
- Unless the expanded water is returned to the supply pipe in line with Water Regulation requirements i.e. it should not be capable of being drawn off as drinking water, this is how most systems below 15 litres capacity work, it must have a means of absorbing the expansion within the system. This is usually either:
  - An expansion vessel; or
  - Contain an integral expansion system
- A means to prevent back flow contamination, a check valve or non-return valve usually incorporated within the combination valve
- As the discharge should not connect directly to a drain, a tundish
- Means to avoid waste of water, discharge valves close automatically when pressure or temperature drops

That's the legislation completed, time now to look at the components in more detail.

Unvented hot water storage system components

Remember; as the unvented system is supplied as part of a unit or package the components that form part of the system will be designed to complement one another, so always use the parts designed for the system by the manufacturer.

If incorrect parts were used the manufacturer's warranty would be invalidated.

Let's take a more detailed trip round the system

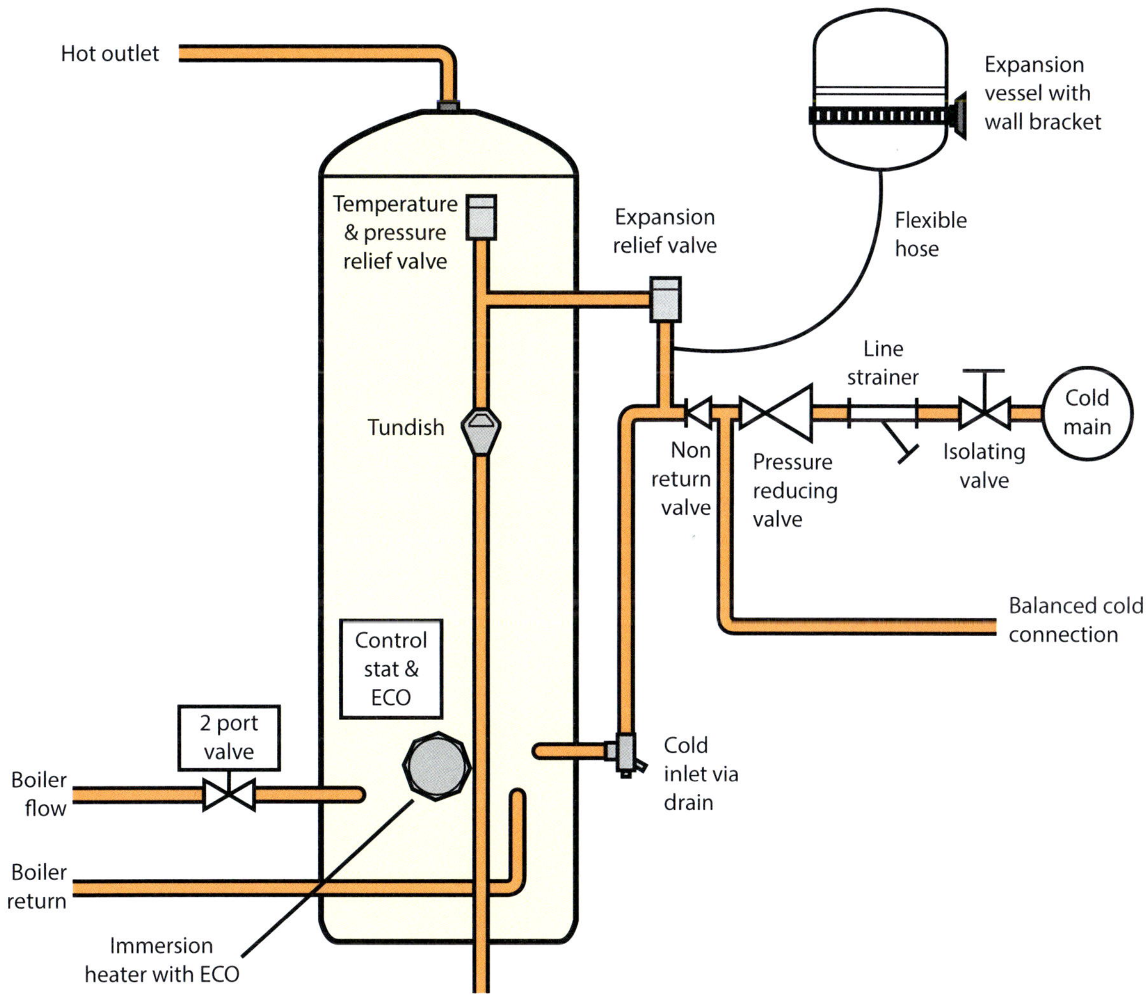

Image 32 **Unvented system components**

This diagram shows the components laid out in the order they are required for clarity and this is the way older systems were assembled, from individual components. Modern systems generally have the strainer, pressure reducing valve, non-return valve and frequently the expansion relief valve assembled into one combination valve.

Let's look at the individual components and then look at some examples of combination type valves.

Storage cylinders

Storage cylinders for unvented hot water storage systems are primarily made out of three types of materials:

- Copper
- Steel (the surface is coated with ceramic, enamel or glass lining material to protect against corrosion)
- Stainless steel

The material from which the cylinder is manufactured will have a bearing on the inlet pressure to which the system is regulated and hence the operating pressure of the system.

For example, copper cylinders are unable to withstand the higher pressures that steel cylinders can (typical system operating pressures are in the range of 1.5 bar – 2.1 bar). The expansion relief valve and the temperature and pressure relief valve will also have lower operating pressures on the copper version.

It is for this reason that the correct manufacturer's replacement parts are always used, using a relief valve with a higher operating pressure could lead to a potentially dangerous rupture of the cylinder.

Coated steel and stainless steel cylinders generally have higher operating pressures.

Detail on the specification for insulation and heat recovery under Part L1 of the Building Regulations is detailed at Module 4.

Isolating (stop) valve

A mains stop valve used to isolate an unvented cylinder, together with its control components, from the supply pipe in order to carry out maintenance etc.

Image 33a **Standard pattern stopcock**

Image 33b **Ball type valve**

Line strainer

This is basically a filter to prevent foreign bodies from passing into the system and thereby to ensure the efficient functioning of controls with moving parts. The line strainer is usually found in the combination valve protecting the pressure reducing cartridge.

Image 34a **Combination valve**

Image 34b **Line strainer in the combination valve**

Pressure control valve

This valve may be either an:

- Inlet pressure reducing valve; or
- A pressure limiting valve

Pressure reducing valves tend to be used more with modern unvented cylinders although you may come across pressure limiting valves when maintaining existing systems. Both types of valve reduce the mains water pressure to the working pressure of the storage cylinder or heater.

The line strainer, check valve and expansion valve may be incorporated in a combination unit with the pressure control valve.

In the event of a malfunctioning pressure control valve, e.g. debris causing an obstruction, an increase of pressure could occur downstream of the valve, which could result in over pressurisation of the unit. This over pressurisation will be relieved through the expansion valve.

Pressure reducing valve (Pressure regulating valve or water governor)

The reducing valve closes by the action of back-pressure in the valve chamber acting across the full diaphragm. This enables it to operate over a wide range of inlet pressures and give close control of outlet pressures from the valve, i.e. the system operating pressure, and optimise performance from the system in service.

They are particularly appropriate for copper cylinders, which operate at lower pressures, e.g. maximum 2.1 bar. Always check out the unvented manufacturer's instructions before installation.

Image 35 **Pressure reducing valve**

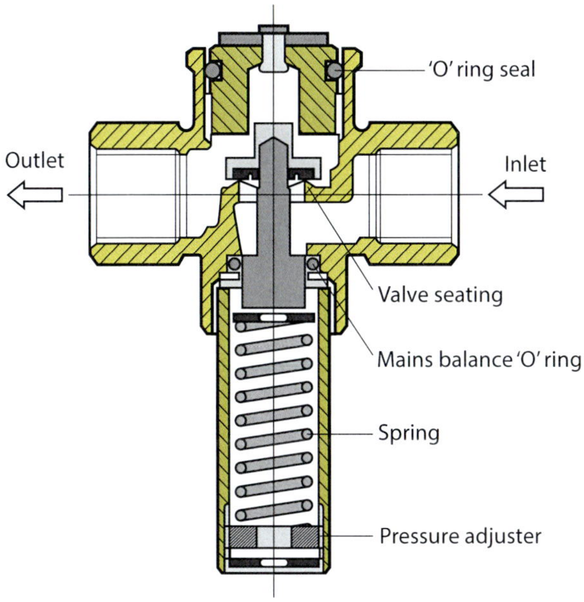

Image 36 **Pressure limiting valve**

Pressure limiting valve

Pressure limiting valves are spring-loaded devices, which operate within specified limits of the set pressure. These limits are in direct proportion to the inlet pressure.

The valve therefore prevents over pressure conditions, but without the close control that is provided by a pressure-reducing valve. If the incoming pressure drops (in peak draw-off times) this permits flow at the pressure remaining in the mains.

Pressure limiting valves were provided with some of the earliest steel unvented cylinders which were supplied in the UK but now tend not to be used. You may find one on a maintenance job.

Non-return/Check valve

A non return (single check) valve is designed to prevent back-flow of water from the system into the supply pipe or mains as the system water is heated up. Image 37 shows a non return/check valve inside a combination valve

Image 37 Non-return/Check valve

Accommodating expansion

There are two primary methods of accommodating expanded water in unvented hot water storage systems – expansion vessels and air gaps.

Use of an expansion vessel

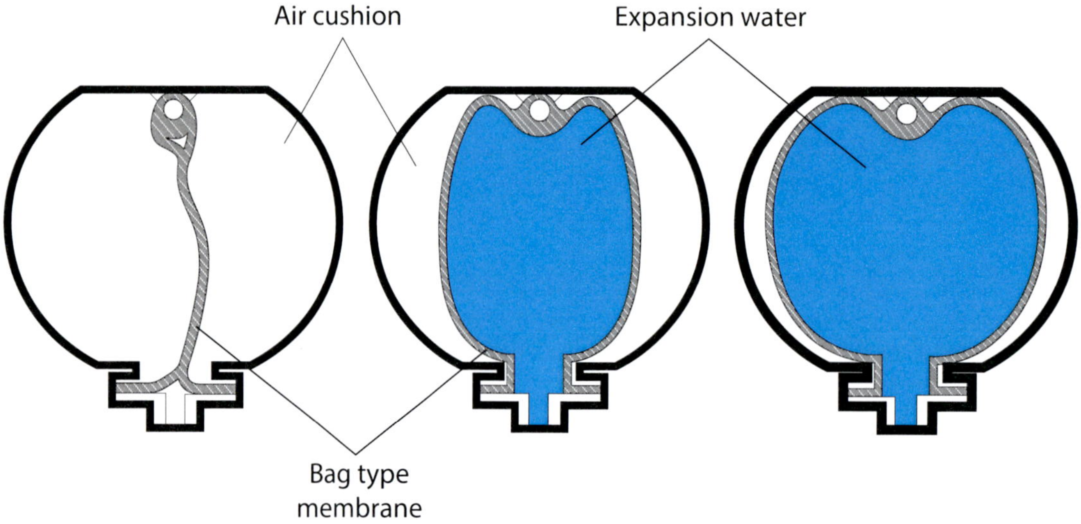

Image 38a Expansion vessel operation

The expansion vessel is designed to prevent wastage of water. It must be capable of accommodating the increase in volume of the expansion water under normal heating conditions, as a minimum, some 4% of the original system contents at cold start up. Remember the reference to expansion of water previously?

The charge pressure in the vessel should be the normal operating pressure recommended by the cylinder manufacturer. The size of the vessel and its charge pressure will usually be pre-determined by the manufacturer of the system. Some of the early expansion vessels were made from mild steel and the pre-charge gas used was nitrogen to prevent corrosion of the vessel.

The vessel must be fitted after the check valve in order to allow expansion. If it was connected before the check valve no expansion could take place. It is usually fitted to the cold inlet pipework to prevent deterioration of the elastomeric membrane as contact with heated water reduces its life expectancy; this may also prevent a build up of scale in the connection to the vessel in hard water areas.

Some manufacturer's designs mean the vessel is connected directly to the cylinder, some to the combination valve and some provide purpose-made fittings for connecting the vessel. Always check the manufacturer's instructions.

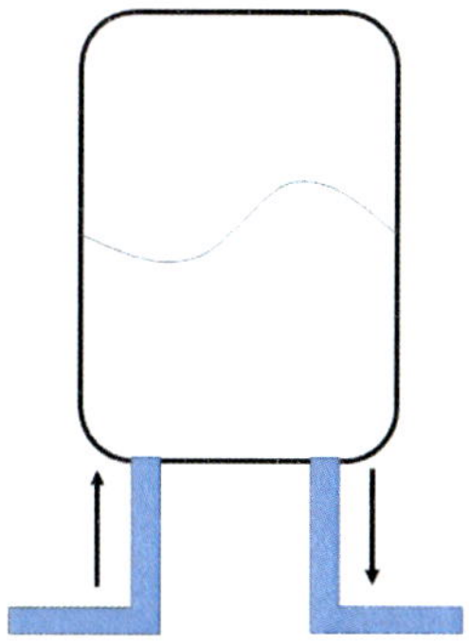

Image 38b Flow through expansion vessel

Water Regulations recommend that a 'flow through' type of expansion vessel is used in order to avoid a deadleg connection with stagnant water in it. This type of vessel has two connections or a split manifold connection so fresh water passes through the vessel.

Although recommended they are rarely seen in domestic installations.

Here are a few examples

A B

Image 39 a and b External expansion vessels

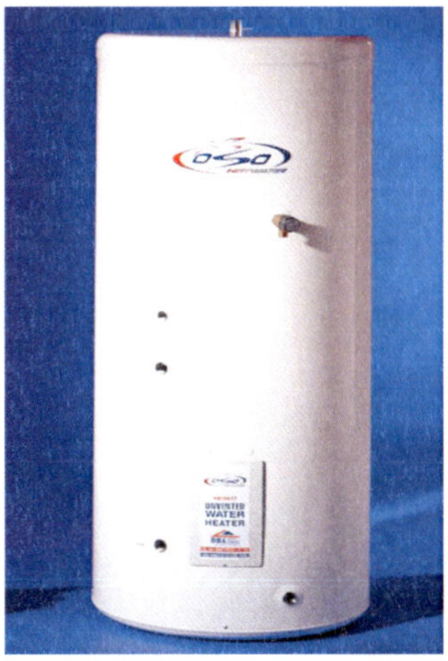

A B

Image 40 a and b Internal Expansion Volumes

Cylinder air gap (Bubble top)

Not all cylinders require a separate expansion vessel, some use an air gap at the top of the storage area to accommodate the expansion of the heated water. The air gap is created during the commissioning of the cylinder as a result of the draw-off connection for hot water having a "dip pipe" which reaches down from the top of the cylinder into the stored water.

Because the water can only be drawn off via the dip pipe a large "bubble" of air is trapped in the top of the cylinder which can absorb the expansion.

The surface of the water will normally be turbulent when heating up, (think of the water in a saucepan on a hob) this can lead to the water gradually absorbing the air trapped in the top of the cylinder. One manufacturer has patented a baffle to prolong the life of the 'bubble'. It is a good idea to re-instate the 'bubble' when carrying out regular servicing.

When commissioning or re-commissioning a "bubble-top" always follow the manufacturer's instructions in order to ensure the expansion volume is correct.

Expansion valves

These valves can be either 'lever-top' pattern (operates by lifting a lever) or 'twist-top' pattern as shown in the photographs below left, and bottom right. Below right is a combined expansion and pressure reducing valve.

A

B

Image 41 a and b **Expansion valves**

The expansion valve automatically discharges water when the pressure within a system has reached a predetermined level.

The two main features of the valve are to protect the system in the event of:

- Failure of the expansion vessel or in the case of a 'bubble top cylinder' a loss of expansion volume, i.e. loss of air pressure causes an imbalance of pressure between water and air in the device.
- Failure of the pressure reducing valve to maintain the inlet cold water pressure at the design operating pressure of the system.

The expansion vessel will open in the event of a failure of the expansion vessel, bubble or pressure reducing valve.

Image 42 **Combined expansion and pressure reducing valve**

The pressure at which the valve begins to operate is pre-determined by the cylinder Manufacturer. On modern systems, expansion valves typically tend to operate at pressures ranging from 6 to 8 bar. With earlier systems the operating pressure for the expansion valve was usually 0.5 bar higher than the system operating pressure; for copper cylinders it is usually 3.5 bar.

No other valve should be fitted between the expansion valve and the storage cylinder. The valve is pre-set by the appliance manufacturer and must not be adjusted.

Temperature relief valve (or combined temperature and pressure relief valve)

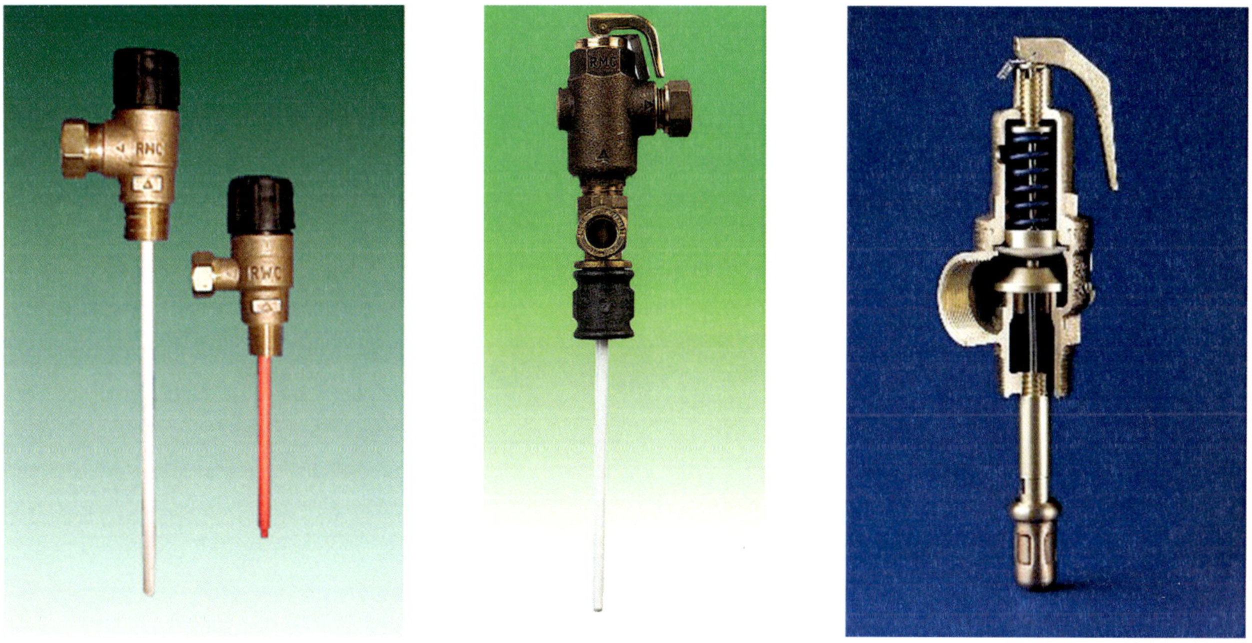

A B C

Image 43 a, b and c Temperature relief and combined temperature/pressure relief valves

In order to protect the system and for the safety of the occupants, a temperature relief or combined temperature and pressure relief valve is provided. The valve is usually set to fully open at a maximum temperature of 90 – 95°C, depending on the manufacturer but in all cases must have a maximum temperature of 95°C.

This valve is supplied and fitted to the storage vessel and is adjusted to open at the set temperature by the manufacturer. The method of valve operation varies depending on the manufacturer.

The main principle however, is that the sensing rod establishes the water temperature in the cylinder; at excessive temperatures the valve begins to open discharging the over-heated water to an appropriate termination point. The water in the cylinder is subsequently replenished with cold water from the supply pipe thus cooling the cylinder.

Temperature relief valves are designed to incorporate a pressure relief function, the pressure relief function is operates at a higher pressure than that of the expansion valve.

The use of a combined temperature/pressure relief valve does not exclude the need for a separate expansion valve. If the pressure relief function operates it may not re-seat, so that it may be necessary to replace the whole temperature relief valve after rectifying the fault condition that caused the over pressurisation.

An anti vacuum valve was often built into some early temperature relief valves and fitted on cylinders in the UK market. It is no longer considered necessary for steel cylinders but is still incorporated into relief valves fitted to copper cylinders.

High limit thermostat (non self resetting thermal cut out)

Direct cylinders will have their high limit thermostat built into the immersion heater as discussed earlier. Depending on manufacturer, the immersion heater will be screwed into the top of the cylinder or about 200mm up from the base of the cylinder.

Indirect cylinders must have a high limit thermostat often referred to as an energy cut-out (ECO). It must not be adjustable or be able to be reset without manual intervention (non self-resetting).

The actual temperature at which it cuts off the energy supply to the cylinder is fixed by the manufacturer and is usually in the range 80 – 85°C but in all cases must prevent the temperature of the stored water exceeding 90°C.

The ECO is normally combined with the control thermostat and will usually be fitted into a preformed dry phial pocket in the cylinder. Some manufacturers build the control thermostat and ECO into the cylinder as seen earlier on page 11.

The combined thermostat/ECO shown in Images 44a to c has twin phials, one for each function, fits into a dry pocket and has the reset button under a screw-on cover.

Image 44a

Image 44b

Image 44c

If the cylinder is also connected to an alternative heat source such as a solar thermal hot water system in a twin coil cylinder, then a separate ECO must fitted to prevent overheating by the solar energy. The ECO's can be identified by the fact that the lower one will be for the solar thermal and the one for the primary heating system will be higher, approximately half way up the cylinder.

Control thermostat

The control thermostat, whether for directly or indirectly heated cylinders, is adjustable and should normally be set at a temperature which will deliver water at 60°C at the tap 30 seconds after fully opening it as recommended in BS EN 806 Part 2. Normally for domestic systems this will be at 60°C as this will prevent the proliferation of *Legionella* bacteria. Although they are adjustable you should recommend to the customer that they are left at the recommended settings.

Vacuum relief valve (anti vacuum valve)

As mentioned earlier in this module, on early light gauge copper cylinders manufactured for the UK market, a vacuum relief valve was required to equalise the pressure within a storage vessel, should it be subjected to a vacuum condition. This would normally be incorporated within the temperature or expansion relief valve. The one being indicated here is part of an expansion relief valve.

Image 45 **Vacuum relief valve on pressure relief valve**

In an unvented domestic hot water system there are some situations where a partial vacuum may occur:

i. When a cylinder is drained down without venting the system, e.g. via a draw-off tap.

ii. When water is drawn off faster than the inlet can replace it. The problem is more likely to occur if the draw-off points are below the base of the unit particularly where mains pressure is relatively low, e.g. 2 bar or less, or where the supply is restricted due to an inadequately sized main supply pipe.

iii. When a hot system cools after the closure of the water supply isolating valve.

For light gauge copper cylinders designed for open vented application a reduction of pressure, below the external (atmospheric) pressure, can be sufficient to cause the cylinder to collapse or implode.

Modern high pressure heavy gauge copper and stainless steel cylinders designed for unvented hot water systems, are far less susceptible to implosion; so vacuum relief valves tend not to be supplied with unvented hot water storage systems.

Tundish

Image 46 **The tundish**

Both the expansion valve and the temperature relief valve are designed to discharge water if a fault occurs in the system.

A blocked discharge pipe could affect the correct operation of either temperature relief or expansion valves, or both if their outlets are joined together, to avoid this; an air break is formed in the discharge pipe by means of a tundish.

The tundish, currently supplied manufactured from plastic, must be fitted in the same space as the cylinder and safety device. The inlet and outlet of the tundish must be separated by a AUK3 type air gap. This is a minimum of 20mm or twice the diameter of the inlet pipe, whichever is the greater. Most inlet connections are 15mm so this will mean that a 30mm air gap is required.

We have already covered the requirements of the pipework connection to the tundish, we will cover the discharge pipe requirements from the tundish in more detail later in Module 2.

Combination valves

New unvented hot water storage systems tend to be supplied with combination control valves (for ease of connection) containing four or more of the functions we have described previously, as shown next.

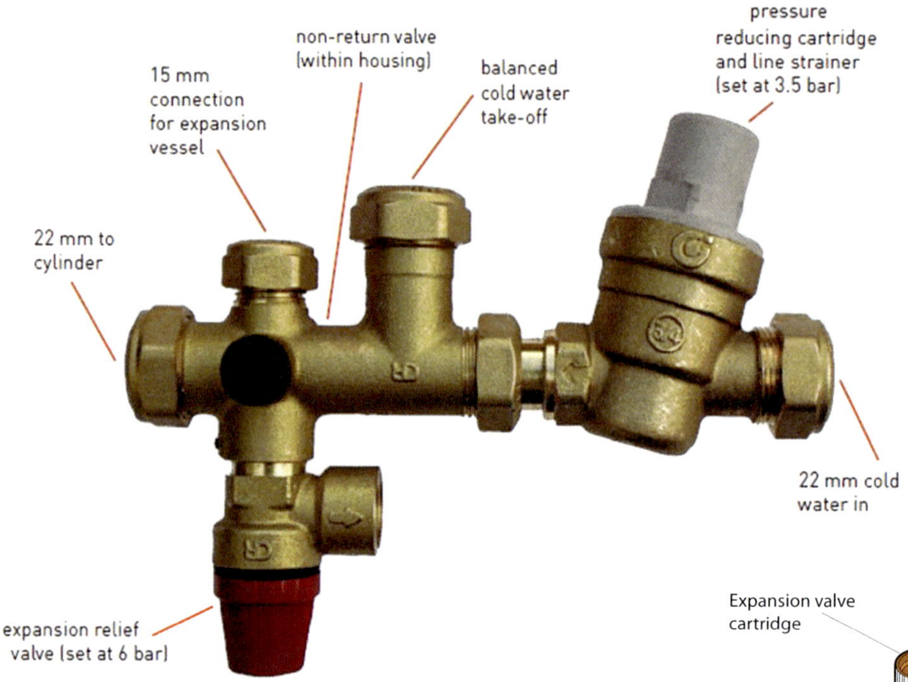

Image 47 Combination valve

Although Image 47 shows the pressure reducing cartridge as set to 3.5 bar it may well be a different pressure depending on the type and manufacturer of the cylinder. The normal range is 2.1 to 3.5 bar and care must be taken to select the correct model when replacing a defective valve.

Right: breakdown of a combination valve with an isolation valve

Previously, most cylinders were made of copper. With the onset of unvented heaters, however, it was necessary to increase the thickness of the copper to 1.6mm or grade 2, which made the copper cylinders heavier and more costly.

Some unvented cylinders are made from mild steel, which allows for high pressures but requires a lining to separate the steel from the water and reduce reaction between the two; this is often enamel or glass.

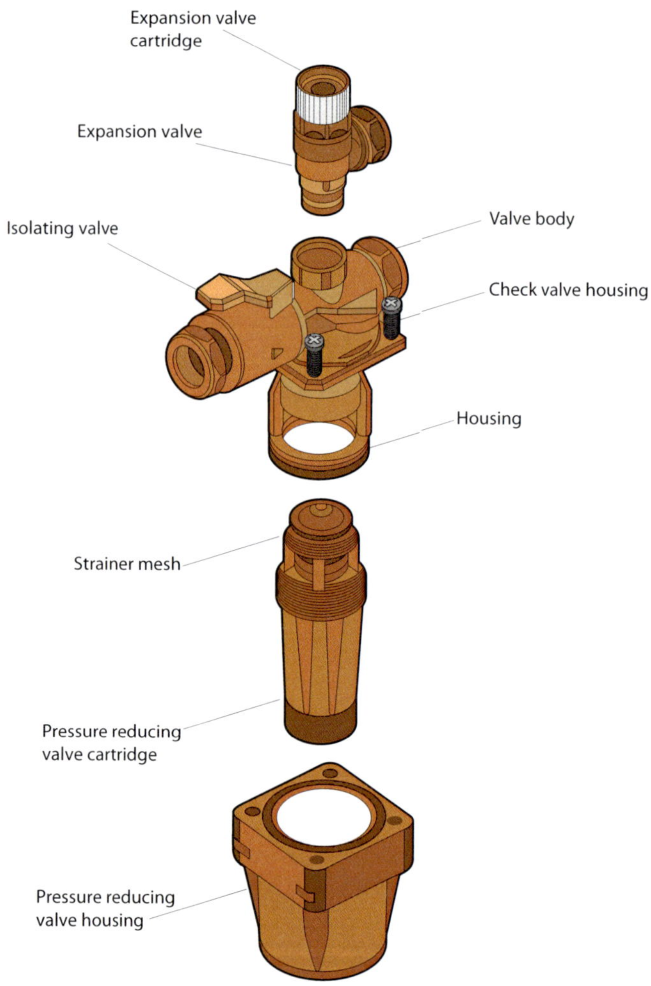

Image 48 Combination valve with isolation valve (exploded view)

The cylinder requires a sacrificial anode or similar device to deal with possible reactions from imperfections in the lining between the two elements.

Most anodes are shaped like a rod and made from an alloy that consists of magnesium fitted in the top of the cylinder. In order to protect the cylinder from corrosion the anode erodes and the molecules then settle onto the imperfections on the surface of the cylinder preventing further corrosion. The anode must be inspected annually for decay and replaced if necessary.

The images below illustrate:

- A new anode and one that has been in service for some time and has eroded
- An articulated replacement anode that is segmented like a chain. The linked sections allow you to bend the rod for installation.

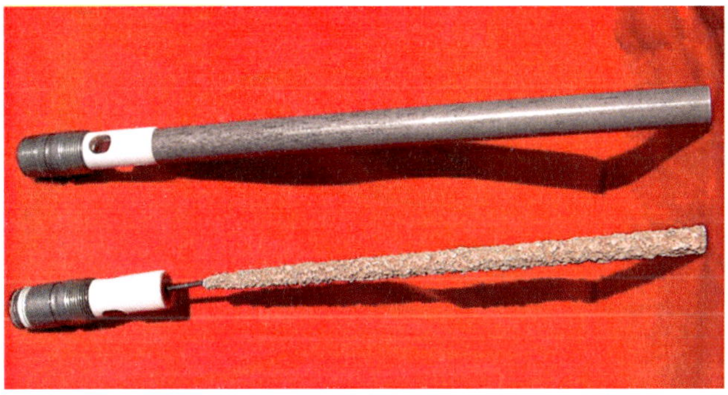

Image 49 **New anode and an eroded anode**

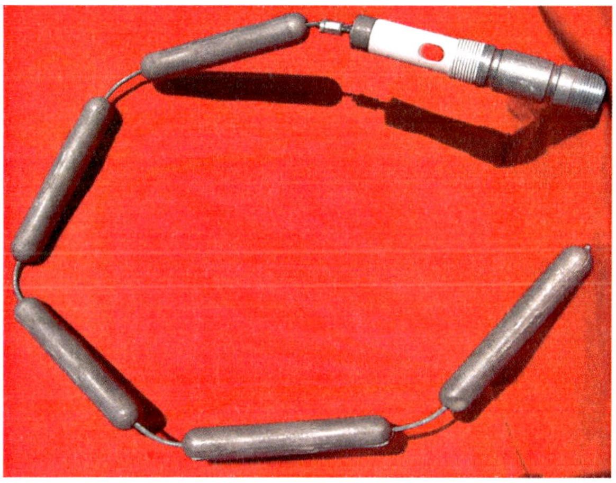

Image 50 **An articulated replacement anode**

Local water authorities distribute water supplies between different areas of the UK, resulting in areas that were previously free of corrosion now receiving more acidic water and hence have problems of corrosion.

The problem of corrosion is an important consideration when specifying a water heater that uses anodes.

Well that's it for this module you should know all the basics of an unvented hot water storage system and the legislation that goes with it, just one task for you before we move on and that's a few self assessment questions, just to see how you are getting on.

Try and do the questions without peeking at the answers!

Self assessment questions

1. The Building Regulations require controls to be fitted on unvented hot water storage systems to prevent the water exceeding what temperature?

2. What is the purpose of an expansion valve which is installed in an unvented hot water storage system?

3. According to the Building Regulations what is the maximum permitted distance between the outlet of a temperature relief valve and its connection to the tundish?

4. State one method of accommodating expanded water in an unvented hot water storage system.

5. What is the purpose of a pressure reducing valve in an unvented hot water storage system?

6. At what unvented system storage volume do the Building Regulations begin to apply?

Self assessment answers

1) 100°C

2) To relieve excess pressure in the system should the expansion device or pressure reducing/limiting valve become defective

3) 600mm

4) Air gap in cylinder (bubble top) or expansion vessel

5) To regulate the water pressure entering the system and hence control the operating pressure of the system

6) 15 litres

Module 2

Design and Installation of Vented and Unvented Hot Water Systems

Contents

Aim of the module

On completion of this module and following attendance on the BPEC Vented and Unvented Hot Water Systems Training Course you should be able to:

- Identify the key issues affecting the design and installation of unvented hot water systems
- State the typical commissioning procedures for vented and unvented hot water systems

The next module will build on your learning from Module 1, next we'll look at a number of essential design and installation issues which you need to consider when working on hot water storage systems and in particular the unvented variety.

Design and installation of unvented hot water storage systems

When installing an unvented hot water system several factors need to be considered in terms of design principles, these include:

- Incoming water supply pressure and flow rate
- Size of unvented cylinder
- Pipe sizing
- Discharge pipe sizing and positioning
- Multiple cylinder installations
- Secondary circulation systems
- Balanced or unbalance supply
- Safety factor
- Location of cylinder
- Unvented hot water storage systems and uncontrollable heat sources
- Control systems

Let's look at these in a bit more detail.

Incoming water supply pressure and flow rate

Unvented hot water storage systems are designed to provide better water flow rate at terminal fittings (taps and showers) simultaneously, always assuming that there is sufficient water pressure and flow rate from the mains.

So what do you need to take into account?

OFWAT has a declared standard of service indicator called DG2; this requires water companies to deliver a supply of water at 1 bar at a flow of 9 litres per minute at the external stop tap. This does not override the duty to supply water constantly at a pressure to reach the upper floors of properties.

An unvented cylinder depends on the pressure and flow provided by the mains service, and a pressure of 1 bar and flow rate of 9 litres per minute is not likely to result in an effective delivery.

It is therefore essential to establish whether the incoming supply to the property is capable of providing the type of service promised by the installation of a new unvented cylinder.

Although there are 'rule of thumb' figures for what flow you will get from a 15mm copper pipe or 25mm MDPE pipe you must assume nothing and check the actual delivery.

The first items to check are the standing and working pressures, these can be tested by connecting a pressure gauge to a cold water mains fed tap and opening the tap.

This will show the standing pressure and if you open another cold mains fed tap this will show the working pressure.

Image 51a **Poor standing pressure**

Image 51b **Good standing pressure**

To check the available flow rate simply use a flow cup or weir gauge whilst measuring the working pressure.

Image 52 **Poor flow**

Image 53 **Good flow**

Image 54 **Good flow and good working pressure**

After recording the readings, check the manufacturer's published information for the cylinder you are considering installing to confirm that the cold supply is capable of providing an adequate performance.

What to if the water supply is not good enough?

There are a number of options for overcoming supply problems, but each will have a significant cost attached to it. Advise the customer before committing to an order as the customer may not want to proceed.

Option 1

The method that has been around for a number of years is designed to overcome poor inlet pressure by the use of a boosted cold pressure set. This is usually a small cold water cistern (break tank) which is supplied normally via a float valve and the outlet of the cistern is connected to a pump.

There is also a pressure switch connected to the outlet connection which brings on the pump when the supply pressure to the taps and hot water drops to a pre-determined level, and thus provides a much better service at the terminal fittings.

The installation of the cistern and pump must comply with the Water Regulations 1999.

Option 2

A recent introduction is a booster pump designed for use mainly in open vented systems but which may be fitted in line in the cold supply of the property **if the local water supplier will allow it**.

The normal Water Regulations apply and the flow rate must not exceed 12 litres per minute. The pump is designed to boost the pressure by around 0.5 to 0.75 bar and has a built in flow switch which detects when a tap has been turned on and starts the pump.

Option 3

The third option for boosting the flow is by means of an accumulator; this is like a large expansion vessel which fills with water at higher pressure when demand is low i.e. during the night and then releases that pressure into the cold and hot supplies when needed. There are numerous accumulator systems each working or connected in slightly different ways depending on the manufacturer's design. Some are a combination of booster set and accumulator. If in doubt which is best for your installation, consult the manufacturers.

Schematic diagram of typical SuperStream system

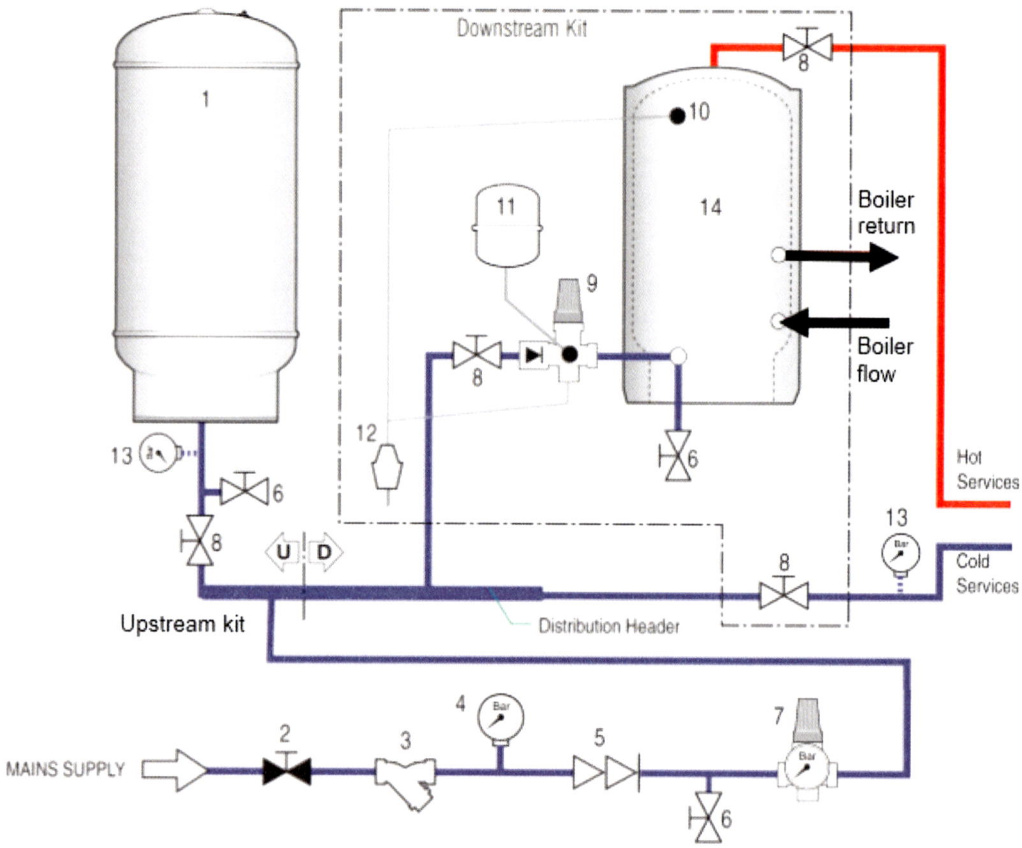

1. SuperStream Accumulator Vessel
2. Stop Cock (not supplied)
3. Inline Strainer
4. Pressure Gauge
5. Double Check Valve
6. Drain Cock (not supplied)
7. Pressure Reducing Valve c/w gauge
8. Full Bore Isolation Valve
9. Multibloc valve
10. Temperature & Pressure Relief Valve
11. Expansion vessel (not supplied)
12. Tundish
13. Extra gauge locations
14. Oso SuperStream Unvented cylinder

Image 55 Using an accumulator

The rule of thumb recommended minimum delivery for an unvented system is 1.5 bar at a flow of 20 litres per minute. Most customers expect and will get a better service than this with good design and even suggested improvements by you. If the minimum delivery is less than this you should consult the manufacturers for advice.

Size of unvented cylinder

There are two categories when sizing a cylinder:

- Domestic
- Commercial/light industrial.

The size of the property and the number of bedrooms, baths and showers will determine the size requirements of a domestic cylinder.

An indirect cylinder has a quicker recovery cycle than a cylinder that is directly heated.

What size cylinder do you need?

BS 8558 gives a complicated calculation based on the recovery time of reheating using the varying demand and the input from the boiler/heater $M=(VT)(14.3P)$

Where:

M is the time in minutes taken to reheat

V is the volume of water heated (L);

T is the temperature rise (°C);

P is the rate of heat input to water (kW).

P uses:
3kW for an immersion heater, 6kW, 10kW and 15kW for small, medium and large boilers respectively.

The difficulty here is that you have to make so many assumptions as to how the end user will draw off the hot water – it makes it almost impossible to calculate accurately.

The old BS 6700 or CIBSE Guide figure of 35 to 45 litres per person per day still works well when sizing a hot water cylinder.

This manufacturer's selection chart shows they are using similar figures.

Table 1 Manufacturer's Cylinder sizing guidance

Which cylinder to use		
Application	**Indirect**	**Direct**
Bedsit	70l	70D
1 bed 1 bath & shower	125l	125DD/145DD
2 bed 1 bath & shower	125l	145DD/170DD
3 bed 1 bath & shower	145l	210DD
4 bed 1 bath & shower	170l	210DD
4/5 bed 2 bath & shower	210l	250DD
4/5 bed 3 bath & shower	250l	300DD
Light commercial uses	250l	250DDD/250DDD 300DDD/300DDD
Other commercial uses	300l	250DDD/250DDD 300DDD/300DDD

A bedsit is likely to be a single occupier so the manufacturer's smallest cylinder is fine

A 1 bed apartment may well have 2 occupants so we go up a size

A 2 bed dwelling could be 2 or 3 occupants so 3 x 35 to 45 litres is a minimum of 105 and a maximum of 135 litres

And so on...

Sizing for a commercial or industrial application

The commercial/light industrial market has a far greater range of demands than the domestic market; hence it is necessary to size the hot water storage capacity by looking at the individual sites' requirements for hot water.

The capacities stated in the table 2 are the minimum storage requirements for commercial and light industrial cylinders. The total storage requirements can be adjusted if the site allows for recovery time between multiple uses or if the demand is over a longer time period.

Table 2 Commercial and industrial cylinder

Application	Hot water storage
Hand washing	1 litre per person per use
Hand washing (spray tap)	0.5 litres per person per use
Showering (general)	10 litres per minute per shower
Showering (sports)	6 litres per person per 5 minute shower
Bath	55 litres per bath
Kitchen	3-5 litres per meal served
Dishwasher	2.5 litres per meal served
Washing machine	20 litres per wash
Cleaner's sink	10 litres per sink
Hair dressing	10 litres per shampoo

Having decided on the size of cylinder required the next item to check is its rating under the Energy Related Products Directive (ERP). Most people are familiar with the colourful A to G energy labelling of domestic products such as washing machines, fridges and driers. Since 26 September 2015 new boilers, heat pumps, and water heaters need to display a label to show their energy performance.

The labels follow the same standardised format already established for white goods. For simple hot water cylinders and tanks including indirectly heated cylinders, the ratings are based on the standing heat loss of the cylinders, and display the capacity of the cylinder and its heat loss in Watts. The A to G rating is based on the relationship between these two figures.

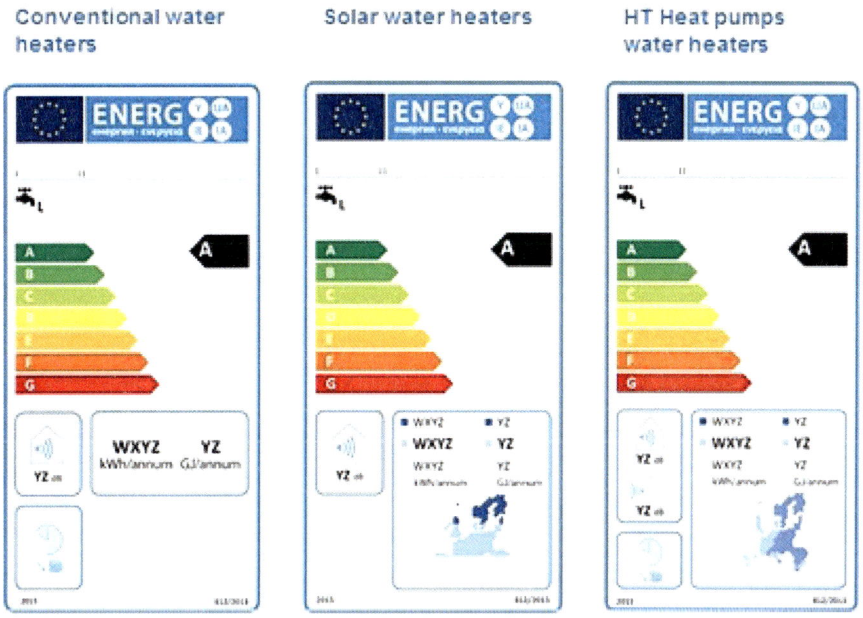

Image 56 **Cylinder Energy Labelling**

There are separate labels for solar systems and heat pumps. Higher labelling categories such as "A+" and "A++" will be reserved for appliances using renewable or low carbon inputs such as solar thermal or heat pumps.

If the cylinder is installed as part of a heating system then the whole system will need a 'package' label showing the overall rating of the system based on the boiler, cylinder and controls installed.

Most boiler manufacturers have label creation tools on their websites or visit http://www.beama.org.uk/tools.html.

Having selected the cylinder the following points need to be considered:

- The hot water pipe runs should be kept as short as possible and any dead legs should be removed
- There must be adequate clearance around the cylinder to facilitate installation, maintenance and repairs
- The floor where the cylinder is located should be of substantial construction to safely carry a full cylinder (1 litre of water equals 1kg) and be able to withstand potential minor leakages
- The length of the D2 pipe to the safe discharge point should be considered. Alternatively, when connecting to an internal soil stack system incorporating a waterless trap, the length of the pipe run to the central discharge stack must be considered
- If the unvented hot water cylinder is located in a cellar/basement, a self-priming pump must be installed in a sump to lift the discharged water to the safe discharge point.

See pages 68 to 75 for design and sizing of the D2 pipework and use of a discharge pump.

Pipe sizing

As with cylinder sizing, pipe sizing is best carried out following established design procedures in order to ensure that correct flow rates are achieved at terminal fittings in accordance with BS EN 806 Part 2.

BS EN 806 Part 3 contains a simplified pipe sizing method based on Loading Units (LU) each LU equates to a flow rate of 0.1 l/s. The standard also states that nations can use their own references for flow rates etc. when calculating loading units for pipe sizes.

BS 8558 contains the following Table 5 for use in calculating LUs in the UK.

Table 5 **Design flow rates to outlet fitting and appliances** [A), B)]

Outlet fitting or appliance	Rate of flow L/s	
	Design rate	Minimum rate
WC cistern (to fill in two minutes)	0.13	0.05
WC flushing trough (per WC served) [C)]	0.15	0.10
Urinal cistern (each position served)	0.004	0.002
Washbasin	0.15	0.10
Handbasin (spray or mixer taps)	0.05	0.03
Bidet	0.10	0.10
Bath (G ¾)	0.50	0.20
Shower head (15 mm)	0.20	0.10
Kitchen sink (G ½)	0.20	0.10
Kitchen sink (G ¾)	0.30	0.20
Washing machine	0.20	0.15
Dish-washing machine [D)]	0.15	0.10

A) These figures can be used for sizing each individual pipe to the appliance referred to.

B) The choice of outlet fittings/appliances can affect the flow rate delivered at the outlet fitting/appliance. Manufacturer's advice should be sought.

C) WC flushing troughs are recommended where anticipated use of WCs is more frequent than once per minute.

D) The manufacturer should be consulted for required flow rates and pressures to washing and dish-washing machines for other than single dwellings.

Using the flow rates and LUs on this example will allow you to quickly size the pipework.

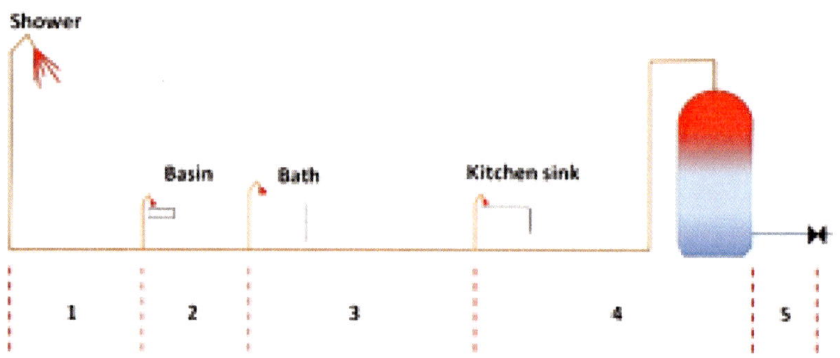

Section 1, shower only

From table 5 in BS 8558 Flow rate – 0.2 l/s =2 LU

Table 3 — Loading units for determination of pipe diameters

Table 3.1 — Hot-dip galvanised steel								
Max. load	**LU**	**6**	**16**	**40**	**160**	**300**	**600**	**1 600**
Highest value	LU	4	15					
DN		**15**	**20**	**25**	**32**	**40**	**50**	**65**
d_i	mm	16	21,6	27,2	35,9	41,8	53	68,8
Max length of pipe	m	10	6					

Table 3.2 — Copper														
Max. load	**LU**	**1**	**2**	**3**	**3**	**4**	**6**	**10**	**20**	**50**	**165**	**430**	**1 050**	**2 100**
Highest value	LU			2			4	5	8					
d_a x s	mm	**12** x 1,0			**15** x 1,0			**18** x 1,0	**22** x 1,0	**28** x 1,5	**35** x 1,5	**42** x 1,5	**54** x 2	**76,1** x 2
d_i	mm	10,0			13,0			16,0	20,0	25	32	39	50	72,1
Max length of pipe	m	20	7	5	15	9	7							

From this extract from Table 3 in BS EN 806-3, 2 LU can use 12mm copper for a maximum pipe run of 7 metres. The full table also allows other options such as polybutylene, PEX and polypropylene.

Section 2, shower plus basin – **0.2 + 0.15 l/s = 3.5 LU**

From Table 3 in BS EN 806-3, 3.5 LU needs 15mm copper or 16mm PE-X or PB

Section 3, shower plus basin plus bath – **0.2 + 0.15 + 0.5 l/s = 8.5 LU**

From Table 3 in BS EN 806-3, 8.5 LU needs 18mm copper or 20mm PE-X or PB

18mm is not a commonly used size in the UK so 22mm would be used

Sections 4 and 5 will need to carry the full load so – **0.2 + 0.15 + 0.5 + 0.2 = 1.05 l/s = 10.5 LU**

From Table 3 in BS EN 806-3, 10.5 LU needs 22mm copper or 25mm PE-X or P

Discharge pipe sizing and positioning

One of the key factors to take into account when designing an unvented hot water storage system is the siting of the cylinder and its control components so that it will allow for safe termination of the discharge pipe. Hence, the distance from the cylinder to the discharge pipe termination point is important and will determine where the cylinder is located. **Remember, it is vital to comply with Building Regulations and manufacturer's instructions.**

Before we look at discharge pipe sizing we need to look at safe termination points. The Approved Document to the Regulations divides the discharge pipe into two sections D1 between the safety valve and the tundish and D2 between the tundish and the termination point.

Approved Document G3

The Approved Document G3 states that:

- Each of the temperature and expansion relief valves or combined temperature and pressure relief valves should discharge either directly or by way of a manifold via a short length of metal pipe (D1) to a tundish. Some manufacturers provide a flexible connection from the valve to a manifold.
- The diameter of discharge pipe (D1) should be not less than the nominal outlet size of the relief valve.
- The tundish should be vertical, located in the same space as the unvented hot water storage system and be fitted as close as possible to the valve with no more than 600mm of pipe between the valve outlet and the tundish.
- Any discharge should be visible at the tundish. In addition where discharges from safety devices may not be apparent i.e. in dwellings occupied by people with impaired vision or mobility, consideration should be given to the installation of a suitable safety device to warn when discharge takes place e.g. electronically operated.
- The discharge pipe (D2) from the tundish should:
  - have a vertical section of pipe at least 300mm long below the tundish before any elbows or bends in the pipework; and
  - be installed with a continuous fall of at least 1 in 200.
- The discharge pipe (D2) should be made of metal or a material capable of safely withstanding temperatures of the water discharged.
- Where the discharge pipe is connected to a soil stack discharge pipe, it should:
  - contain a mechanical seal, not incorporating a water trap, which allows water into the branch pipe without allowing foul air from the drain to be ventilated through the tundish;
  - be a separate branch pipe with no sanitary appliances connected to it;
  - if plastic pipes are used as branch pipes carrying discharge from a safety device, they should be either polybutalene (PB) or cross linked polyethylene (PE-X) complying with national standards; and
  - be continuously marked with a warning that no sanitary appliances should be connected to the pipe.
- Where a single common discharge pipe serves more than one system, it should be at least one pipe size larger than the largest individual discharge pipe (D2) to be connected.

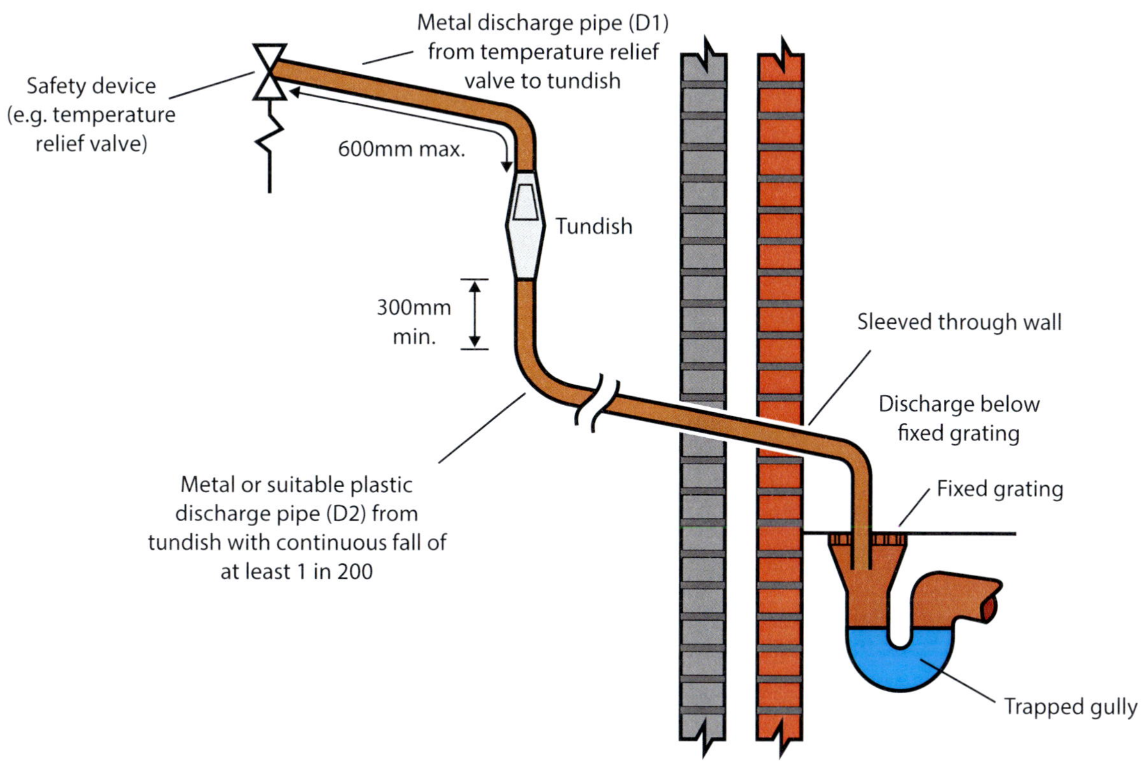

Image 57 Discharge pipes (D1 and D2)

Termination of discharge pipes

The discharge pipe (D2) from the tundish should terminate in a safe place where there is no risk to persons in the vicinity of the discharge.

The following pages show examples of suitable termination arrangements.

Low level discharges

If the low level termination requirements cannot be met and there are concerns regarding the level of risk, the termination should be made into a trapped gully located below a fixed grating and above the water line.

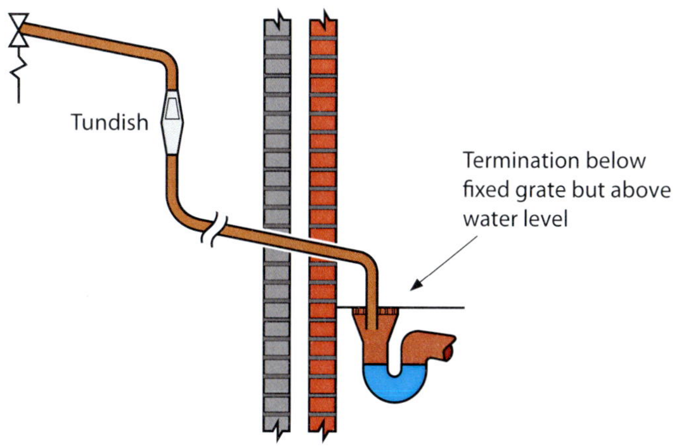

Image 58 Termination into a gully

When discharging into a trapped gully with a fixed grating, as illustrated above the key points to note are as follows:

- A plastic gulley can be used, as long as it is not close to the cylinder
- The discharge pipe must be made of metal
- Exposed discharge pipes should be protected and covered with thermal insulation
- The termination pipe should be fixed tightly, sealed and bracketed
- The loose lid of the trap must be screwed down
- The end of the discharge pipe has to terminate below the lid

- A metal or glazed trap may have to be used

There are two accepted safe low level terminations of the discharge pipe. These are:

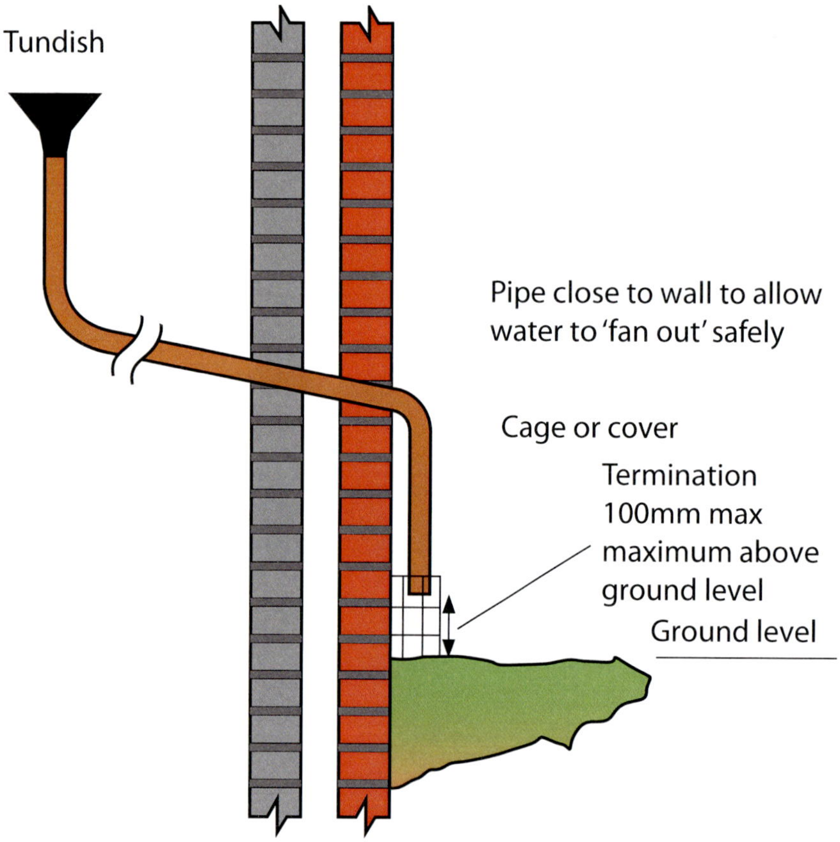

Image 59 Low level termination

When terminating at low level, as illustrated above, the key points to note are:

- The discharge should not be of any risk to persons
- The end of the discharge pipe will require covering with guard or cover that maintains visibility.
- The discharge pipe must be made of metal
- Exposed metal discharge pipes must be covered and protected
- The discharge pipe must pass through a sleeve through the wall
- The discharge pipe must have a continuous fall of 1:200
- The discharge pipe must have a maximum termination point of 100mm from the ground
- The pipe must terminate close to the wall
- The end of the pipe does not have to be visible, provided the tundish is in full view

High level termination

The next three terminations are the only safe and accepted methods for high level termination of the discharge pipe, as identified in the Building Regulations.

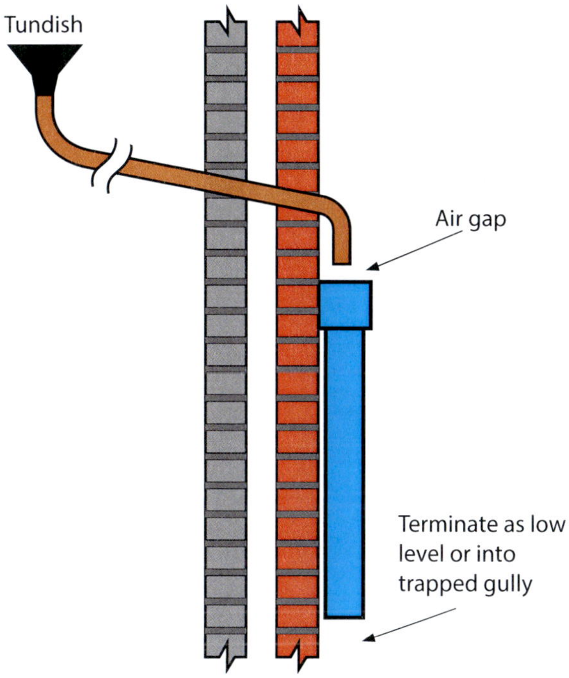

Image 60 **High level (hopper) termination**

When choosing to discharge at high level into a hopper, as shown here, the key points to note are as follows:

- The vertical hopper pipe may be fitted to accept up to six terminations
- An air gap is required at high level for identification purposes (hopper box or additional tundishes)
- The vertical pipe has to be made of metal (D2)
- The minimum size of the down-pipe is one size larger than the biggest D2 pipe going into the hopper
- Safe low level or trapped gully termination is required

The discharge may terminate to a metal hopper, where the end of the discharge pipe is clearly visible (tundish visible or not) or onto a roof capable of withstanding high temperature discharges and which is located minimum 3m away from plastic guttering that would collect such discharges (tundish must be visible).

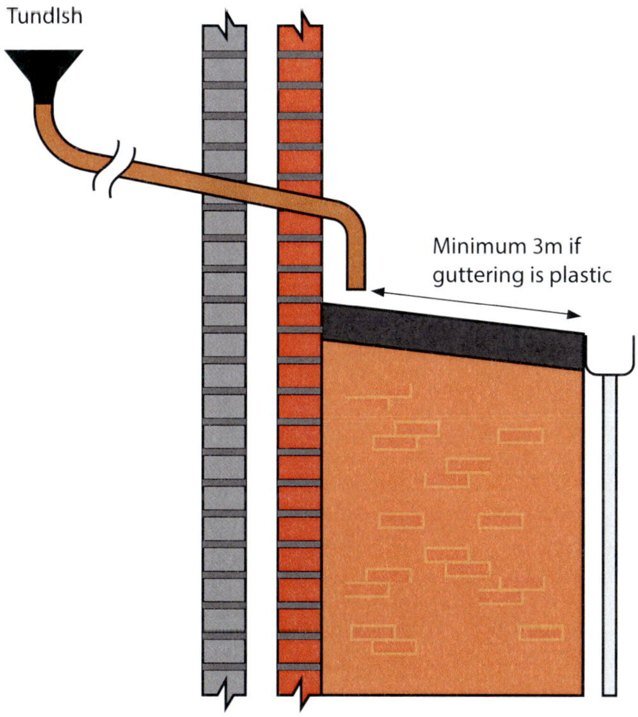

Image 61 **Discharge onto a roof (not to scale!)**

Where the discharge pipe goes onto a roof, as shown left, the key points to note are:

- If termination is into a plastic gutter, the termination of the discharge pipe (D2) must be at least 3m away from the gutter
- If the gutter is metal, there are no minimum measurement requirements
- The end of the pipe must face downwards
- The roof material must be capable of withstanding the high temperatures of the discharge
- It is allowed to discharge onto a flat roof without a gutter
- If the roof is a fire escape route, the discharge pipework must be treated as a low level discharge installation for safety and protection purposes.

Where it is difficult to run the pipework to an external safe discharge point, an alternative method of terminating the D2 discharge pipework internally is acceptable provided the tundish is visible and the other requirements of the Building Regulations are complied with.

The first requirement is that the discharge pipework should not connect to soil stack unless it can be proved that the stack is capable of safely withstanding the temperatures of the discharged water.

If the stack is acceptable then the D2 pipework must contain a waterless mechanical seal to prevent smells, etc., entering the building from the soil stack. A normal water seal trap would dry out quickly.

The D2 discharge pipework must be a separate branch pipe to the soil stack with no sanitary appliances connected to it. If it is made of plastic, it should be of either polybutylene (PB) or cross-linked polyethylene (PE-X) and be permanently marked with a warning that NO SANITARY APPLIANCES should be connected to it.

Installation of the waterless trap and the plastic D2 pipework must be in accordance with the manufacturer's instructions. An example is shown in Image 62, other makes of waterless traps are available.

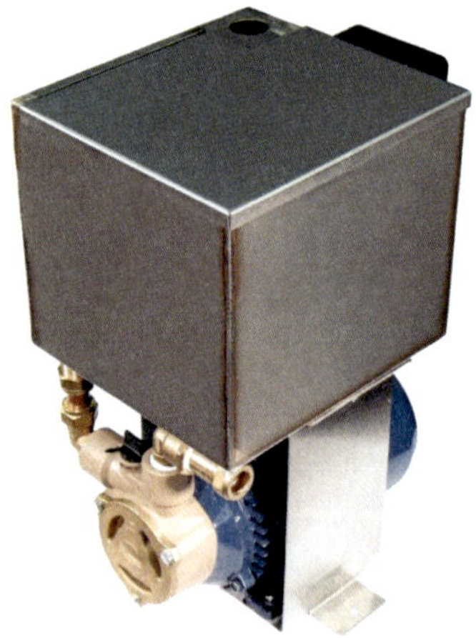

Image 62 **High temperature discharge pump**

Where the only possible location for a cylinder is in a basement or a position where the discharge pipework D2 would have to run upwards above the level of the tundish a high temperature discharge pump will be required.

The pump arrangement consists of a covered metal tank into which the initial section of the D2 pipework terminates. Inside the tank is a float switch which detects the level of water in the tank and starts the pump which lifts the discharged water up to a maximum of 8 metres. When the water level in the tank drops, the pump stops.

The pump can deal with water at up to 100°C so can be fitted to deal with both temperature and expansion relief discharges.

The termination of the D2 discharge pipe from the pump has the same requirements for safe termination as one which came directly from the safety relief valve.

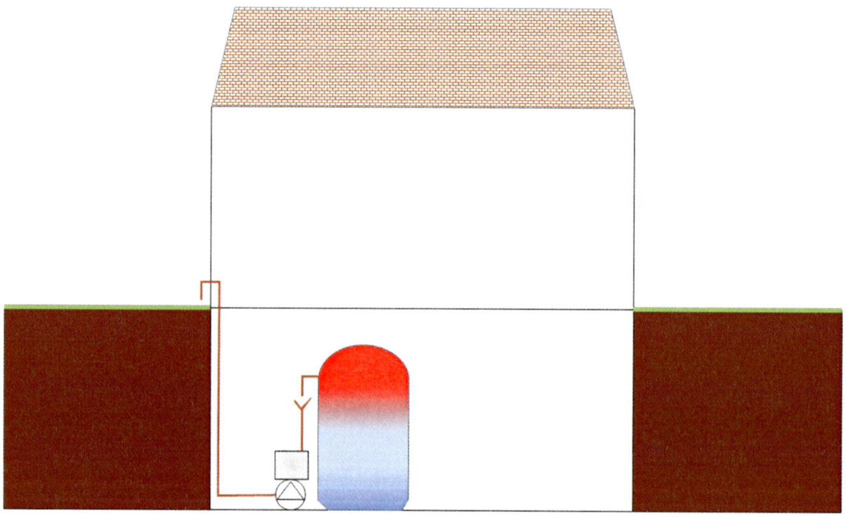

Image 63 **Example of the use of a high temperature discharge pump**

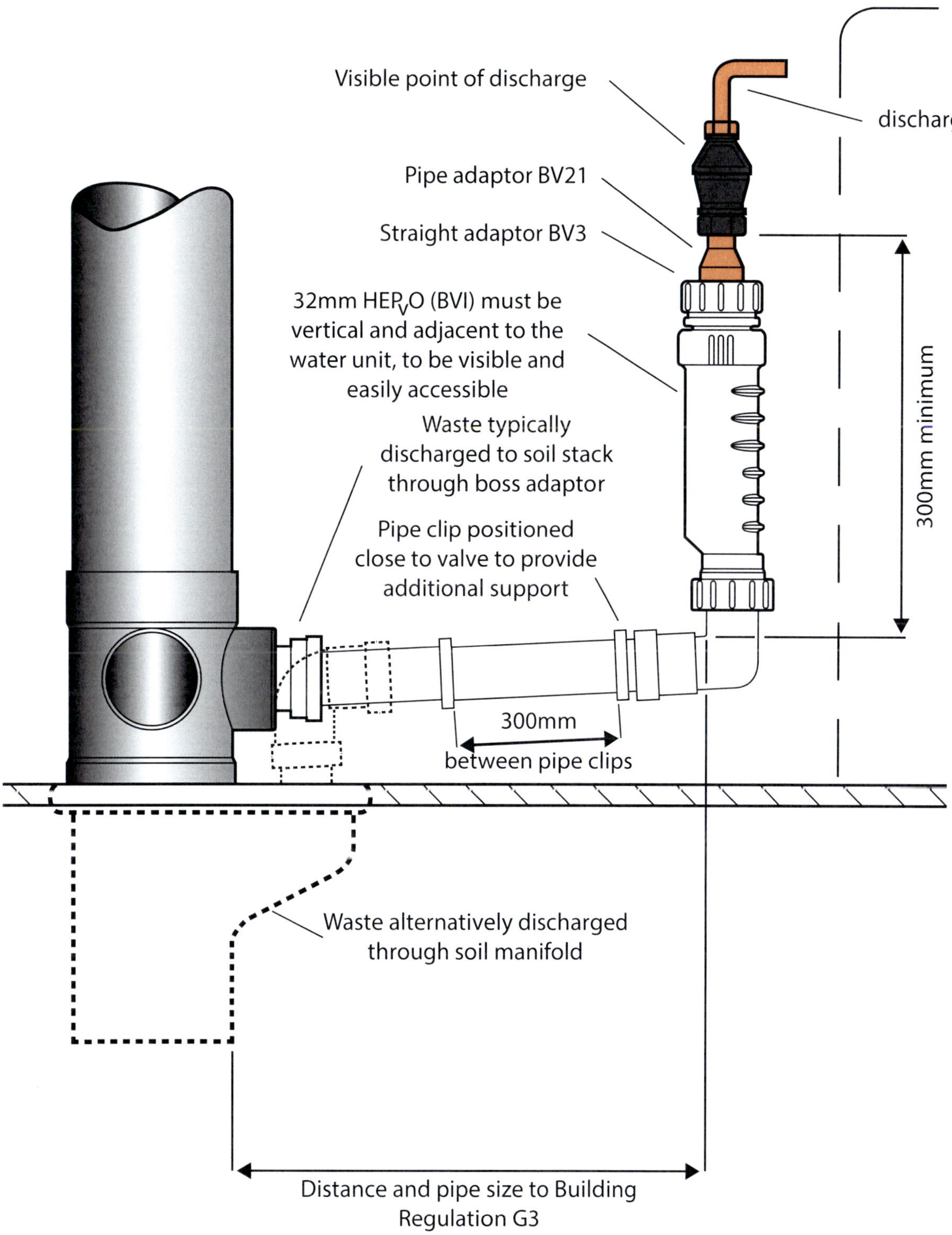

Image 64 Waterless trap method for domestic unvented hot water discharge

Multiple cylinder installations

- Where a single common D2 discharge pipe serves more than one system, for example in a block of flats, it should be at least one size larger than the largest single D2 pipe to be connected.
- If systems are to be installed where discharge is not apparent i.e. in dwellings occupied by the blind etc. consideration should be given to the installation of an electronically operated device to warn when discharge takes place.

Once you have worked out a potential pipe run to a suitable termination point you need to calculate the pipe sizing for the D2 pipework.

We have already seen that the D1 pipework must be a minimum of the same diameter as the outlet of the safety valve it is serving. The basic formula for the D2 pipework is that it must be one size larger than the D1 pipework for maximum runs of 9 metres.

This basic formula is OK for straight pipe runs up to 9 metres long, as soon as a bend or elbow is fitted this adds resistance to the pipe run so the maximum pipe run is reduced.

For example if the D2 discharge pipe is to be run in 22mm pipe and needs to have 2 elbows or bends fitted then each bend or elbow will reduce the maximum pipe run by 0.8 metres so we need to subtract 1.6 metres, meaning maximum pipe run would be 7.4 metres.

If the actual pipe length was 8 metres of 22mm pipe and had bends fitted the equivalent length would be 9.6 metres therefore you would need to increase the minimum pipe diameter by one size, i.e. to 28mm.

The table below shows how to size the D2 pipework.

Valve outlet size	Minimum size of discharge pipe D1	Minimum size of discharge pipe D2 from tundish	Maximum resistance allowed, expressed as a length of straight pipe (i.e. no elbows or bends)	Resistance created by each elbow or bend
G½	15mm	22mm	Up to 9m	0.8m
		28mm	Up to 18m	1.0m
		35mm	Up to 27m	1.4m
G¾	22mm	28mm	Up to 9m	1.0m
		35mm	Up to 18m	1.4m
		42mm	Up to 27m	1.7m
G1	28mm	35mm	Up to 9m	1.4m
		42mm	Up to 18m	1.7m
		54mm	Up to 27m	2.3m

A G½ temperature relief valve has a proposed discharge pipe length (D2) of 6m from the tundish to its termination, the pipe will include 6 elbows.

The calculation is as follows:

Minimum pipe size that could be used is 22mm

From table:

Maximum resistance allowed for a straight length of 22mm copper discharge pipe from a G½ temperature relief valve is 9.0m

Subtract the resistance for 6 number 22mm elbows at 0.8m each = 4.8m

Therefore the maximum permitted length equates to 4.2m

4.2m is less than the actual length of 6m therefore calculate the next largest size

Maximum resistance allowed for a straight length of 28mm copper discharge pipe from a G½ temperature relief valve is 18.0m

Subtract the resistance for 6 number 28mm elbows at 1.0m each = 6m

Therefore the maximum permitted length equates to 12m

As the actual length is 6m, a 28mm (D2) copper pipe will be satisfactory from the tundish to the safe discharge point

Secondary circulation systems

Secondary circulation systems are used to prevent system dead legs. A dead leg on a hot water system, in this context, is an outlet (tap) at the end of a long run of pipework. When the outlet is closed the water in the pipe cools down.

When the outlet is opened, because of the length of the pipe run it takes a long time for the cool water to run out before the hot water appears. This wastes water and energy and on larger systems can lead to Legionella risks.

The Water Regulations provide recommendations for the maximum lengths of un-insulated hot water pipes as follows:

Outside diameter of pipe in mm	Maximum length in metres
12	20
Over 12 and up to and including 22	12
Over 22 and up to and including 28	8
Over 28	3

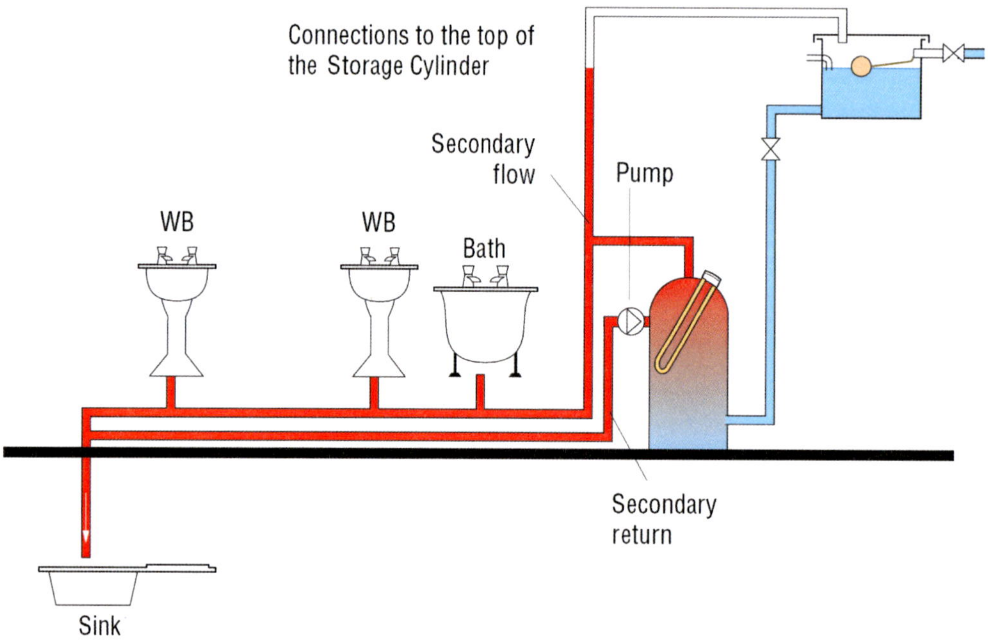

Image 65 Secondary circulation

Secondary circulation systems, in simple terms, consist of a ring main from the hot water draw off which is returned about 25% down the cylinder. The circulation is pumped, and branches are taken off to the various hot water appliances. Another method of tackling the problem of dead legs is with the use of trace heating.

The secondary return pipe normally includes the installation of a non-return (check valve) near the pump, not shown in the diagram. The check valve prevents reverse circulation in the return pipe and the discharge of cooler water when a tap is opened.

The pump is normally operated by a time clock which only permits water to circulate around the circuit when the building is occupied, so conserving energy. Newer systems are installing infra-red proximity sensors so the pump only starts when a person enters a room so saving more energy.

For secondary circulation systems, the secondary return pipe will usually be in 15mm and incorporate a check valve to prevent backflow. A suitable Water Regulations Advisory Service (WRAS) approved bronze circulation pump will be required.

On large systems incorporating either an expansion vessel or cylinder air gap, it may be necessary to fit an additional expansion vessel to the secondary circuit due to the increase in system water content. This should be done if the capacity of the secondary circuit exceeds 10 litres.

Image 66 Spare connection for secondary circulation return

The spare connection on the unvented vessel is for the secondary circulation return.

The table below can be used to determine whether an additional vessel is necessary:

Pipe capacity (copper)

- 15mm o/d = 0.13 l/m (10 litres = 77m)
- 22mm o/d = 0.38 l/m (10 litres = 26m)
- 28mm o/d = 0.55 l/m (10 litres = 18m)

Balanced or unbalanced supply

You need to be careful about where the pressure reducing valve is located in a system. The first image immediately below shows the valve located on the incoming supply, the system having a fully balanced pressure at all outlets. On systems with combination valves the balanced cold supply is usually after the non-return/check valve as shown in Image 67b.

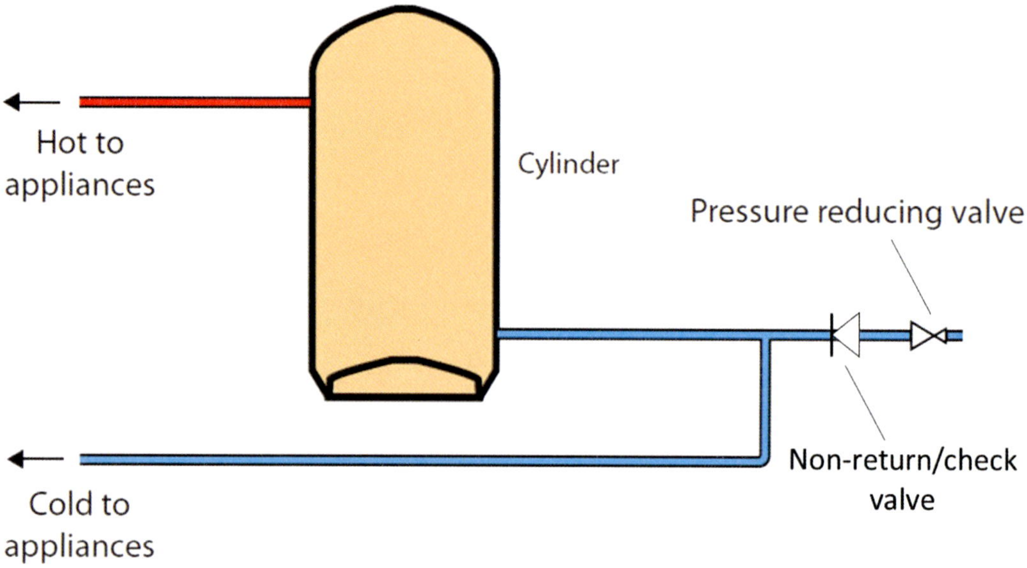

Image 67a Pressure reducing valve on incoming supply

Image 67b Balanced cold connection after non-return valve

The next image below shows an example of a system where the hot and cold water supplies are unbalanced i.e. the cold water and hot water pressures will not be the same due to the positioning of the pressure reducing valve.

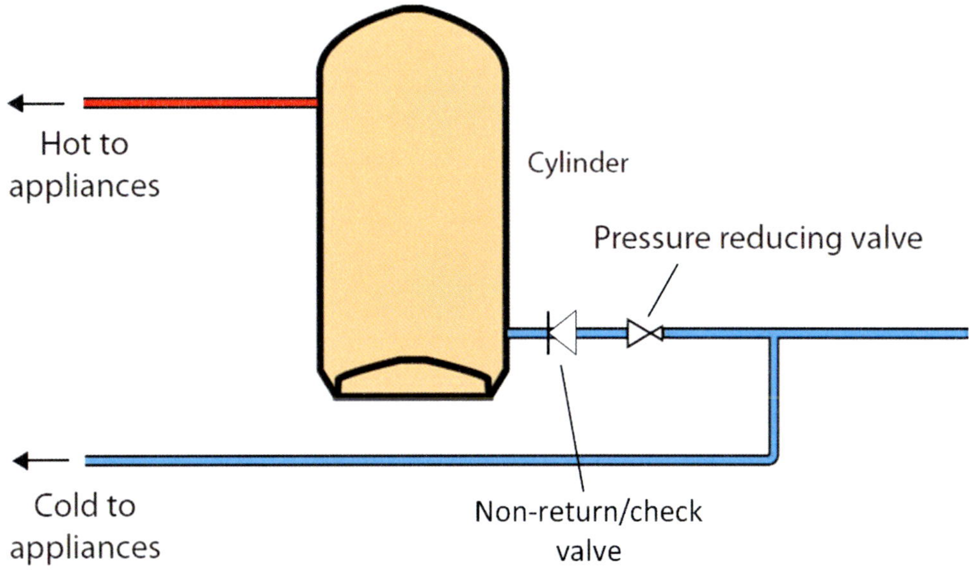

Image 67c **Pressure reducing valve after cold connection**

This image to the right shows the balanced cold connection on a typical combination valve.

It is always preferable to have a balanced pressure system to enable taps to work effectively and usually essential to enable showers to work properly.

Furthermore care must be taken when either carrying out a new installation or converting an existing system to unvented with unbalanced supply pressures that incorporate taps or valves in which the mixing of hot and cold water can occur in the valve body.

In this situation all mixer taps and valves must be fitted with a single check valve on both the hot and cold supply pipes:

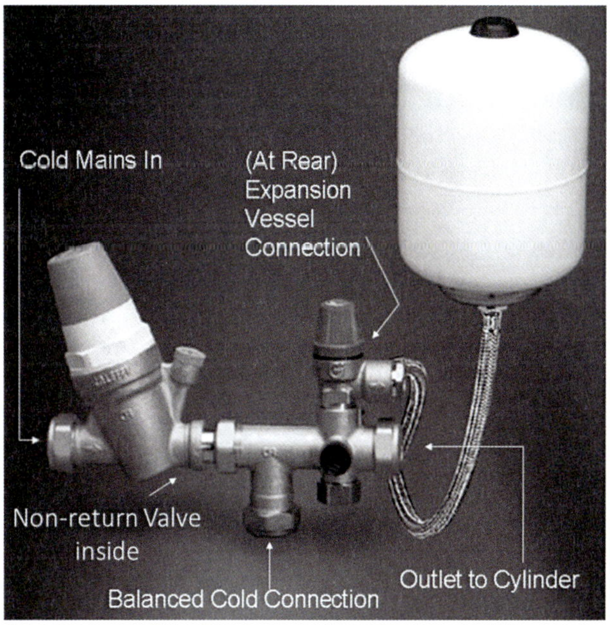

Image 68 **Balanced cold connection on a combination valve**

- To prevent back pressurisation of the cylinder and possible resultant damage of the system or nuisance discharge of the expansion valve, and
- As a legal requirement under the Water Supply (Water Fittings) Regulations 1999 Fluid Category 2 regulations to guard against backflow of water, check valves must be fitted to prevent cross connection and contamination of supply.

The image below illustrates the problem of mixer taps on unbalanced hot and cold supply pressures.

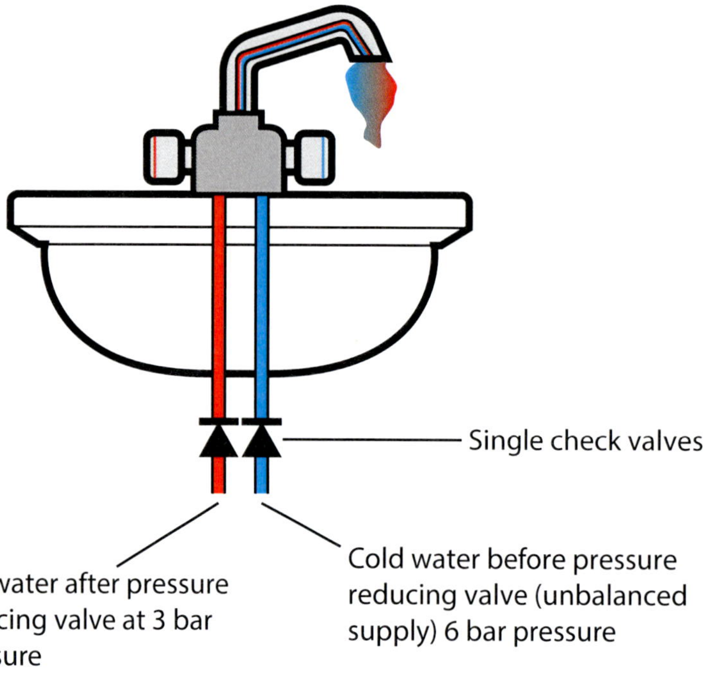

Image 69 **Unbalanced mixer tap supply pressures**

In terms of Image 69, if the check valves were omitted and both taps are opened the cold water supply at higher pressure has the ability to back pressurise the unvented hot water storage system through the hot water pipework.

Combination valving arrangements as shown in Module 1 of this course should be considered when establishing balanced supply pressures, as usually the valve has to be located in the vicinity of the hot water storage cylinder. In order to establish a balanced supply additional piping may prove necessary.

Safety factor

Regulation G3 requires that the hot water supply to any fixed bath must be so designed and installed to incorporate measures to ensure that the temperature of the water that can be delivered to the bath shall not exceed 48°C.

Note: This applies equally to vented systems as well as unvented systems.

This temperature limit can be achieved by use of an in-line blending valve or other appropriate temperature control device such as composite thermostatic mixing valves.

These devices must be used to prevent the risk of scalding, especially to children and older people because of the speed of the hot water emerging from the terminal fittings.

For unvented systems, it is also necessary to check that all hot water distributing pipework and terminal fittings are able to withstand the pressures that will be imposed upon them. It may be possible to fit 'point of use' pressure reducing valves but they may need to have a small expansion vessel fitted too.

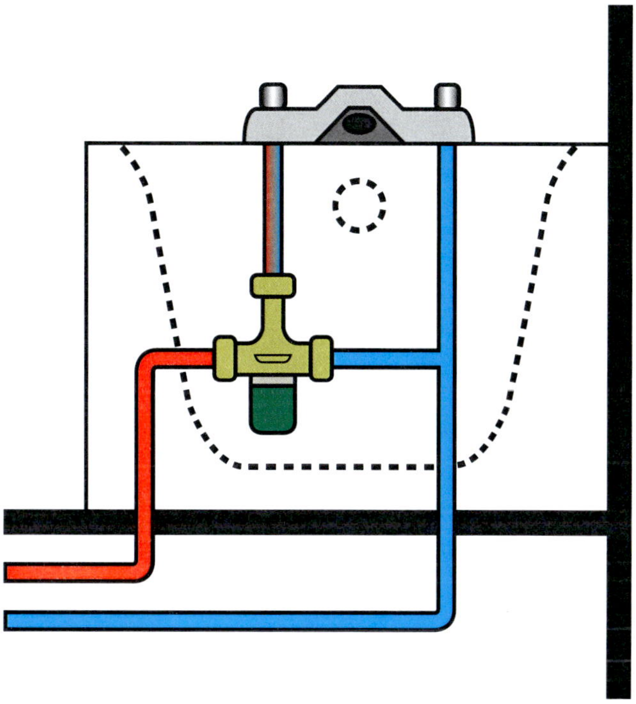

Image 70a **In line blending valve location at the supply to a bath**

Image 70b **In line blending valve in situ at the supply to a wash hand basin**

For general use TMV2 approved valves are acceptable but TMV3 approved valves are required for NHS installations.

Location of the cylinder

When considering the location of a vented or unvented cylinder, the following needs to be considered:

- Sufficient clearance should be provided around the vessel to facilitate maintenance and repairs, details will usually be provided by the manufacturer
- The floor where the vessel is to stand should be of substantial construction to support its weight and the water content. A support for a typical 120 litre unvented unit should be capable of withstanding 170kg.

In addition, the floor surface material should be suitable for purpose i.e. chipboard flooring materials should be in line with Building Regulations and capable of withstanding water permeation.

Hot water storage systems and uncontrollable heat sources

Unvented hot water storage systems should never be supplied from uncontrollable heat sources such as:

- Solid fuel boilers with limited or no control over primary circuit water temperatures
- Solar hot water heating systems with limited or no control over primary circuit water temperatures

Whilst unvented systems should not be supplied from uncontrolled heat sources it is now possible to supply them from solar systems using twin coiled cylinders or even triple coiled cylinders providing that the solar thermal hot water system has full controls and can safely be integrated with the safety controls of the unvented system.

The next two images show a typical example of both a vented and an unvented arrangement for utilising solar hot water heating.

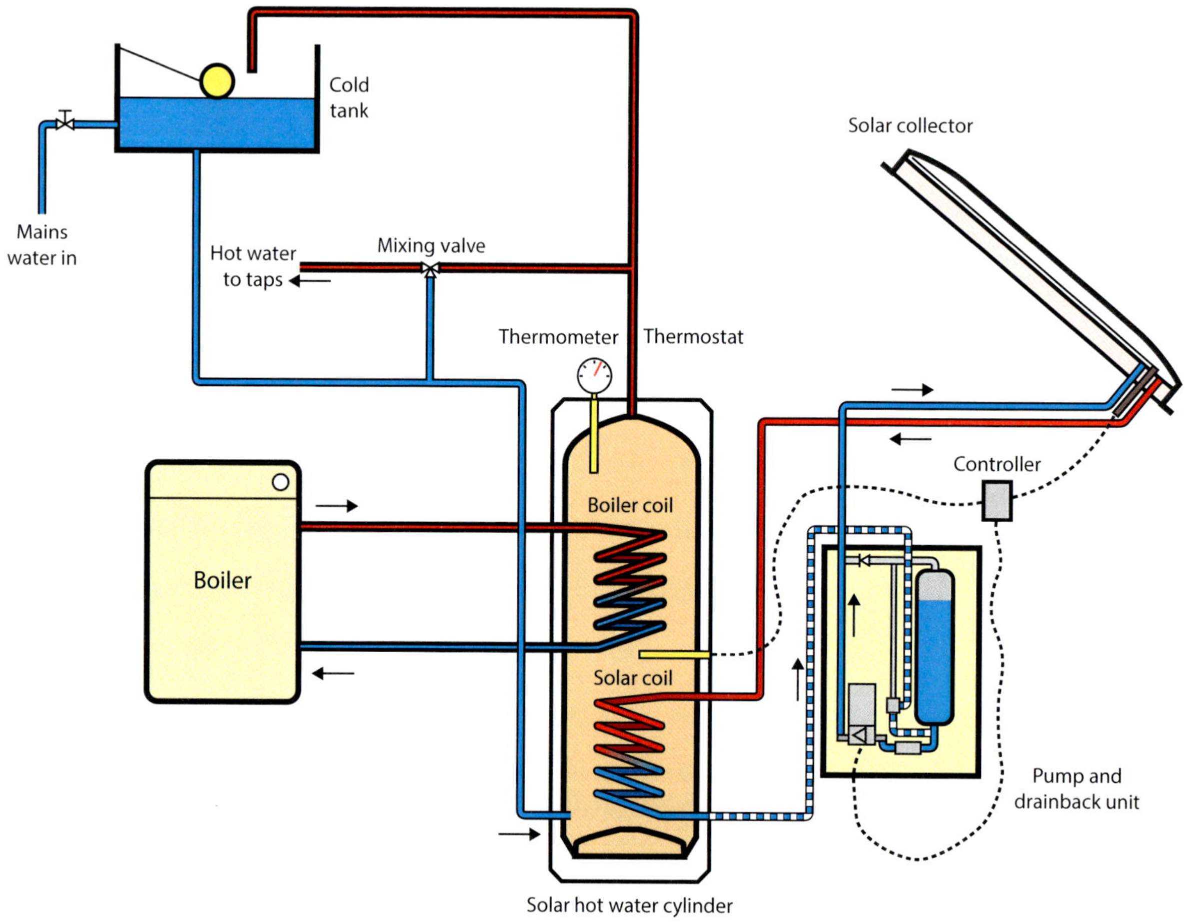

Image 71a Solar hot water heating (vented)

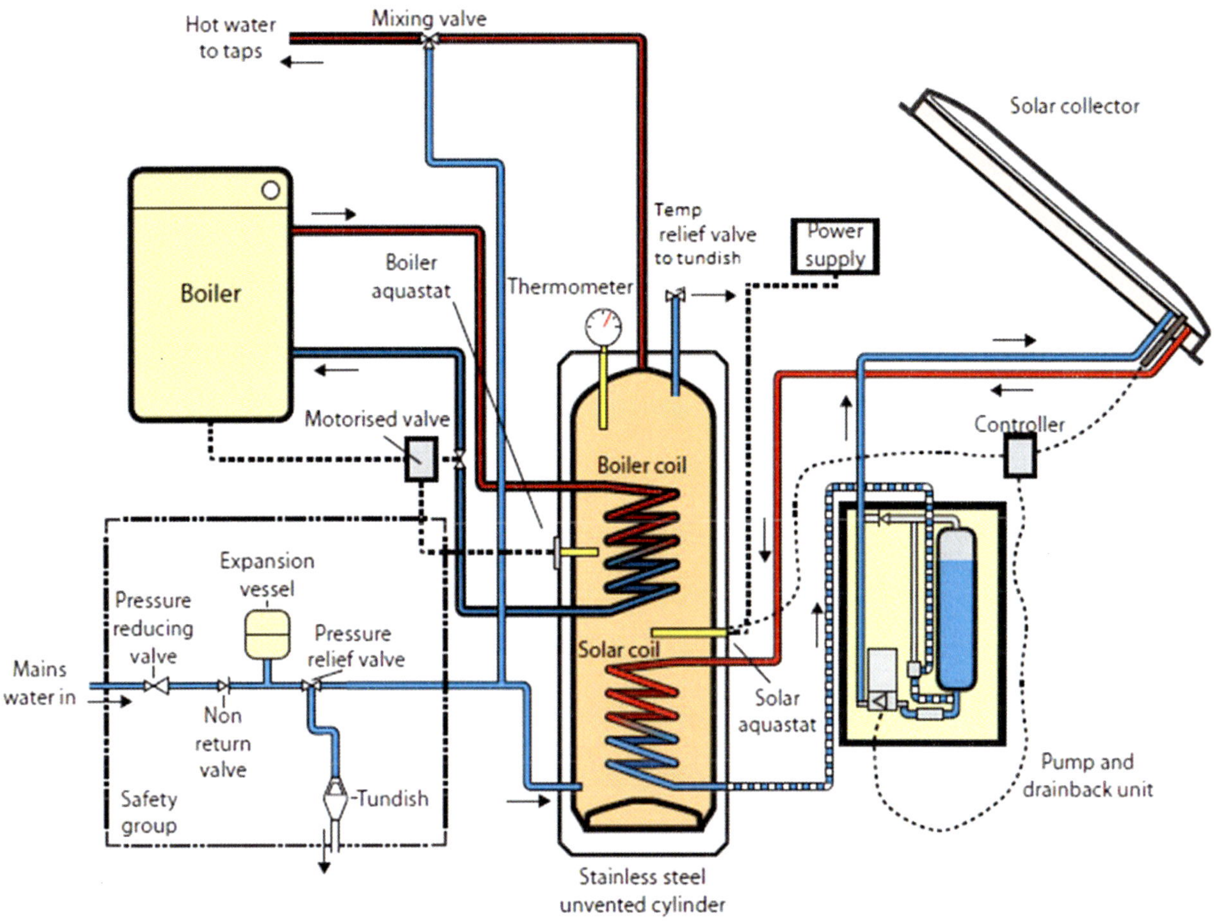

Image 71b Solar hot water heating (unvented)

Unvented cylinders can be installed with solid fuel and biomass heating systems, providing that the:

- Primary circuits are fully pumped
- System has full controls and can safely be integrated with the safety controls of the unvented system
- Manufacturer allows the connection to an unvented system

If you are in any doubt as to the suitability of primary circuit heat source/ system then consult the unvented hot water storage system manufacturer for further guidance.

Control system

Image 72 Motorised valve

You'll probably know that indirectly heated systems must be connected to fully pumped heating systems.

Indirectly heated unvented systems for use in domestic applications require the installation of a motorised valve (usually supplied with the cylinder) to control the temperature of primary water entering the system under normal operating conditions and to isolate the primary heat supply to the cylinder in overheat situations.

The control system can therefore either be based on:

- a 2 two port motorised valve configuration in the 'S Plan', or
- a 3 port motorised valve configuration in the 'Y Plan'

Because the standard 3-port Y plan valve defaults to hot water supply when it is de-energised or faulty, unless advised by the manufacturer, you will be required to install both a 3 port valve and the 2 port valve supplied by the manufacturer. This will ensure that there is effective isolation of the cylinder in an overheat situation. Depending on the type of valve you may also have to order an optional control relay to wire both valves.

The following diagrams show the pipework layout and a functional flow diagram for the electrical circuits when including an additional 2-port motorised valve with a 3-port mid-position valve.

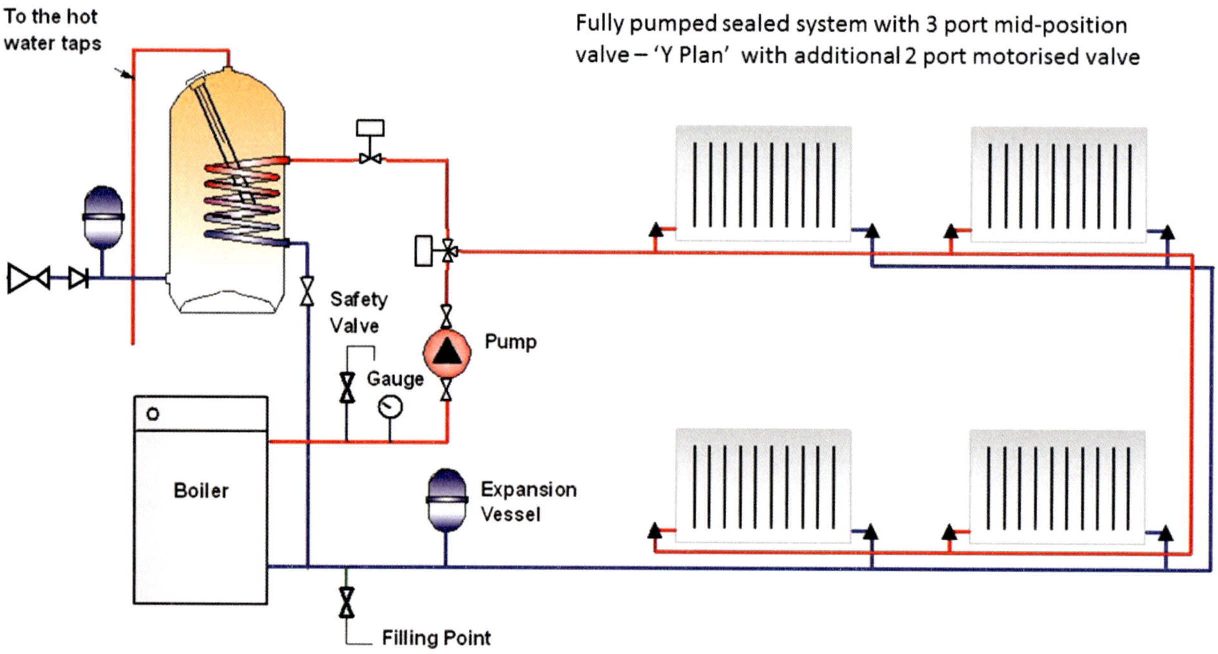

Image 73 Fully pumped Y plan central heating system with additional 2 port valve

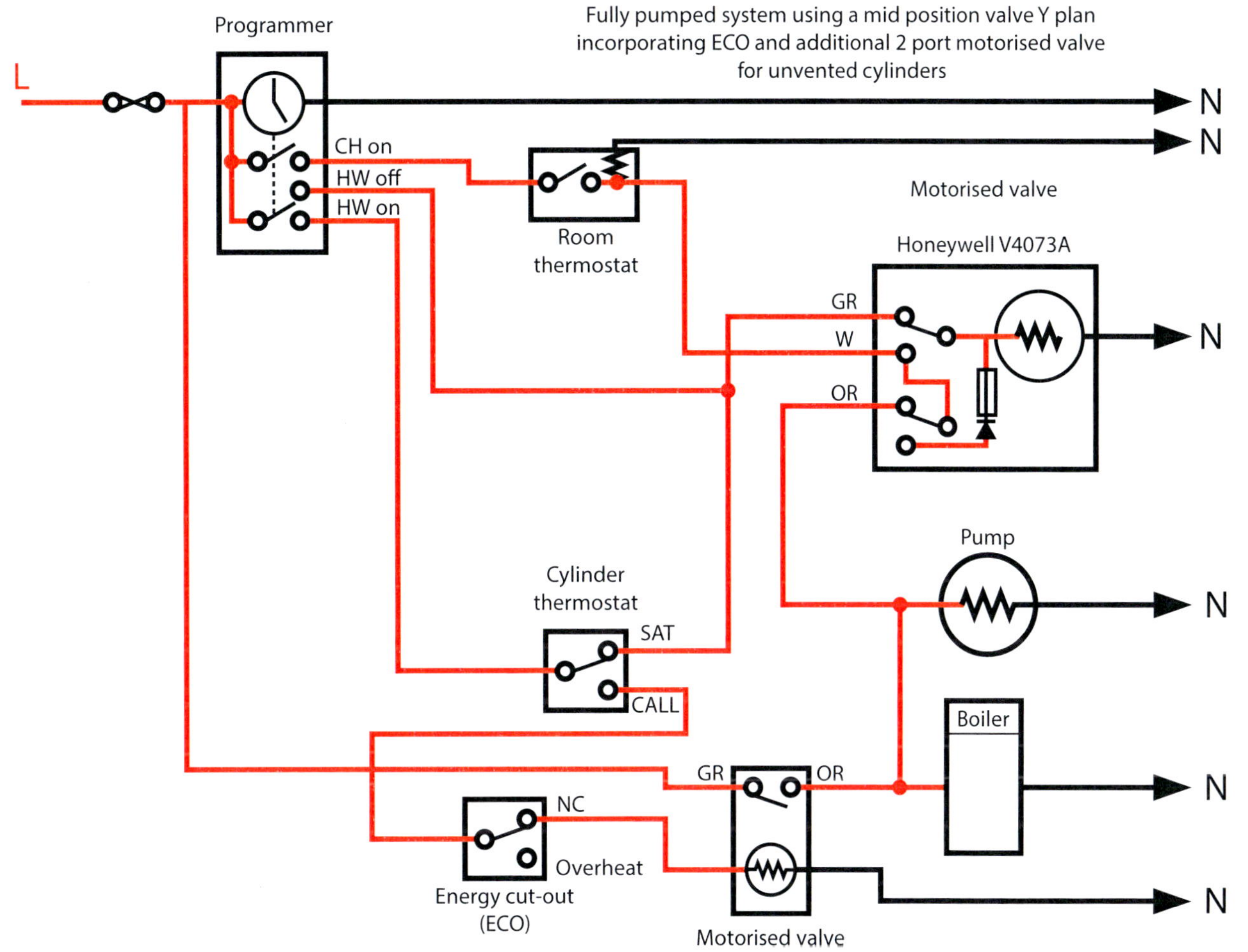

Image 74 Fully pumped Y plan central heating system with additional 2 port valve (wiring diagram)

Electrical connections

The electrical connections to an unvented unit or package should follow the manufacturer's installation instructions and be made in accordance with the latest edition of the IET regulations. The manufacturer of the unit or package will provide detailed guidance on the controls and electrical connection requirements.

This image shows a typical dual thermostat provided with most unvented hot water storage systems, where for ease of wiring both the control thermostat and the thermal cut-out device (high limit thermostat) are housed in one unit.

The manufacturer of the unvented system will provide wiring instructions for the system based on the type of controls being used, you should follow these instructions when dealing with the wiring of systems.

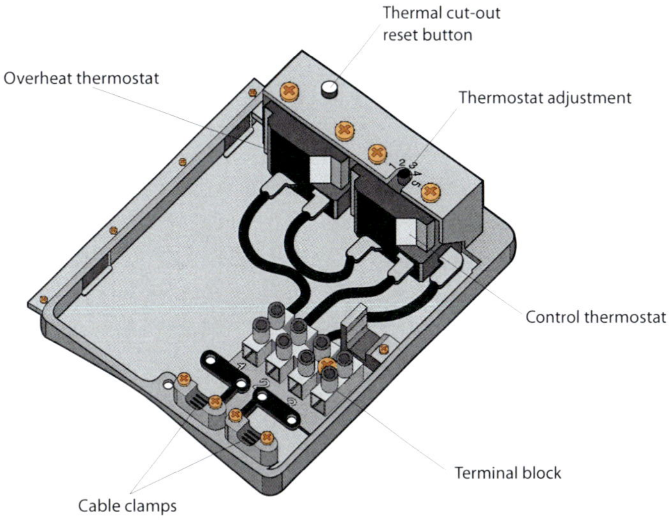

Image 75 Dual thermostat

Commissioning, testing and maintaining open vented hot water systems

Commissioning hot water systems is referred to in Building Regulations L1 and also in the Domestic Heating Compliance Guide. It is important that a structured approach is adopted to commissioning hot water storage systems and in particular that manufacturer's commissioning instructions are followed.

Commissioning can be summarised as:

- Inspecting
- Testing
- And cleansing the system.

Generally, the installer should ensure that:

- Materials and equipment conform to British Standards, Water Regulations and Building Regulations
- Installations are in accordance with the specification
- Installations are in accordance with manufacturer's instructions
- Components operate and function in accordance with manufacturer's instructions
- Then system is water tight
- The system is thoroughly flushed out with drinking water

Commissioning procedure

Commissioning the hot water system should include:

- Making a visual inspection of the system
- Soundness testing (testing for leaks)
- Flushing and disinfection
- Performance testing
- Final checks/handover

Visual inspection

A visual inspection would be carried out prior to filling the hot water system with water. This would include making sure all the fittings/joints had been completed and a double check that the installation complied with all the relevant specifications/regulations. It is also important that visual inspection of pipework and fittings is carried out before any pipework is concealed.

CWSCs and storage vessels should be checked to make sure they are adequately supported and secured, that they are clean and free from swarf and that the cistern is fitted with a cover.

Soundness testing

The system should be filled with water and a visual inspection carried out on all the pipework, components and fittings to check for leaks.

The system should also be pressure tested irrespective of size of installation, or whether it's a new or replacement installation.

BS EN 806 Part 4 sets out the criteria for pressure testing and it differentiates between non-plastic and plastic systems.

BS EN 806 Part 4 should be consulted for more information on pressure testing.

Flushing and disinfection

The system should be flushed out before being brought into operation. This makes sure that any debris, including flux, is removed. It is not necessary to disinfect systems in private dwellings occupied by a single family, but larger installations, say for example in a small hotel would have to be disinfected.

System disinfecting is beyond the scope of this course but training on this topic is available through BPEC.

Performance testing

On a hot water system (depending on the type of installation), performance testing will include:

- Making sure that all the valves and fittings operate correctly
- The water in the CWSC is at the correct level
- Thermostats (boiler, cylinder, thermal cut outs) operate correctly
- Checking that the water temperature in the storage vessel and the terminal fittings (including showers) is as per specification/regulations

Final checks

A final check should be carried out to make sure the fittings and joints have not developed any leaks post pressure testing and that the CWSC and storage vessel are still adequately supported and secured. It's important to replace any pipe and CWSC lagging.

The customer should be given a thorough briefing on how the system works together with any future maintenance requirements and should be given copies of appropriate manufacture's literature.

Commissioning unvented hot water storage systems

The following steps outline the basic commissioning procedures for unvented hot water storage systems; but you should always consult the manufacturer's literature for specific instructions.

- Installation pipe work: Dismantle the line strainer and flush out the pipework thoroughly to BS EN 806 Part 4 standard before reassembling the strainer and connection to cylinder ☑
- Connect to cylinder: Visually check all connections and pipework ☑
- Cylinder thermostat: Check operating temperature set at 60-65ºC ☑
- External expansion: Check pre-charge pressure to manufacturer's instructions ☑
- Filling the system:
    - Close all drain valves and open all hot and cold terminal fittings ☑
    - Open the cold water mains isolating valve slowly and allow the cylinder to fill ☑
    - Allow water to flow from all fittings for a few minutes to flush the system ☑
    - Turn off all fittings and check the system for cold pipework joint security ☑
    - Rectify as necessary ☑
    - By filling up an internal bubble type cylinder, you are setting the initial trapped internal expansion space ☑
- Line strainer: Re-check the line strainer for trapped debris and clean if necessary ☑
- Expansion valve: Slowly turn or lift test lever to check for operation. Hold open for 20-30 seconds ☑
- Temperature and pressure relief valve: Slowly turn or lift test lever to check for operation and ensure that both valves discharge into the tundish. Hold the test lever open for 20-30 seconds ☑
- Tundish: Confirm that the flow from both valves enters and flows through the tundish without spillage, particularly at full discharge flow. There should be no back pressure from the D2 discharge pipework ☑
- Outside termination: Check that the discharged water flows freely from the final termination (compliance with G3) ☑
- Heat up system (direct cylinders):
    - Switch on the system and allow the cylinder to reach its operating temperature ☑
    - Check that the thermostats operate correctly and check stored water for correct temperature ☑
    - Visually, check the hot pipe-work for leaks ☑
    - Confirm that no discharge has occurred during the heat-up cycle from either the expansion valve or the temperature and pressure relief valve ☑

- Heat up system (indirect cylinders):
  - Fill the primary circuit ☑
  - Manually set the 2-port valve to the manual open position, which will fill the primary heating coil. On completion, return the lever to the auto setting ☑
  - Vent any trapped air via the air bleed ☑
  - Switch on the boiler ensuring that the programmer is set to domestic hot water ☑
  - Allow the system to heat up and check that the 2-port valve operates correctly and shuts off the heat source ☑
  - Check that the thermostat has operated correctly (60-65°C setting) by checking the stored water temperature at the nearest hot water tap ☑
  - Visually check for any leakage. Confirm that there has been no discharge from either the expansion or temperature and relief valves during the heat-up cycle ☑
- Benchmark: Complete the Benchmark log book and explain the system to the householder before handing them the Benchmark document together with the commissioning check list and the completed manufacturer's guarantee. Complete the Benchmark sticker and affix to the side of the cylinder ☑
- Insulation: Ensure that all hot and cold pipework connected to the cylinder conforms to Building Regulations Part L pipework insulation requirements ☑
- Completion: Hand over the system and the documentation to the user/owner or the builder/developer ☑

Self assessment questions

1) State two permissible locations for the termination of a discharge pipe from an unvented hot water storage system

a. ______________________________

b. ______________________________

2) It is proposed that an unvented system be installed in a property fed by a supply pipe with a flow rate of 10 l/min and pressure of 10 bar. Will this be suitable?

____

3) A G¾ temperature relief valve is to be served by a discharge pipe 14 metres in length, the pipe will include 2 bends, will a 28mm discharge pipe be suitable?

____

4) A pressure reducing valve is installed on the incoming supply pipe before any other connections, will the terminal fittings be operating under balanced or unbalanced pressures?

____

5) During the filling process for an unvented system should all the terminal fittings be closed, or should some be opened?

____

Self assessment answers

1. Discharge to gully, discharge to low level, discharge into hopper, discharge to soil stack via self-sealing waste valve
2. No – but consult the manufacturer for detailed guidance
3. No – 35mm will be required
4. Balanced pressures
5. One or more terminal fittings should be opened to allow air to vent from the system

Module 3

Maintenance of Vented and Unvented Hot Water Systems

Contents

Aim of the module

On completion of this module and following attendance on the BPEC Vented and Unvented Hot Water Systems Training Course, you should be able to:

- Identify typical vented and unvented hot water storage system maintenance procedures
- Identify a typical range of faults which may occur with unvented hot water storage systems and state the typical fault diagnosis actions to rectify the faults

Maintaining unvented hot water storage systems

Please remember, as we have said before, when replacing any parts on unvented installations, always use those supplied by the manufacturer.

All the components of an unvented hot water system have been approved by an appropriate testing authority such as the BBA, KIWA or WRc and will have been chosen by the manufacturer to require only minimal maintenance.

However, the installer should inform the customer that periodic checks of the equipment are essential for safety and maintenance. The customer should also be encouraged to report any persistent discharge of water through the discharge (drain) pipe to the installer or service contract engineer.

Maintenance and inspection should be carried out to coincide with the annual boiler or water heater maintenance. Experience of local water conditions may indicate that more frequent inspection is desirable, e.g. where water is particularly hard or where the water supply contains a high proportion of solids.

The current Building Regulations Part L1 strongly recommends that the installer or commissioning engineer encourages the owner to continue maintaining the system after the guaranteed period.

You should carry appropriate tools and equipment for testing water pressure, flow rates, temperature of water and to enable reinstatement of any air pressure in an expansion vessel pre-charge.

After an initial visual inspection of a system and discussion with the owner/user about any problems, the following suggested minimum maintenance should be carried out:

1. Check that all existing components are approved for the vessel and are functional.
2. Check to see that all valves are in the correct operating position and all thermostats are properly wired. Be careful with the electrics, as it is common for electricians to replace a failing immersion heater with a standard immersion heater rather than one incorporating the correct thermostats and energy cut-outs. It could be difficult to do this, however, as most unvented hot water storage systems are fitted with irregularly sized components.
3. Check for evidence of recent discharge, such as outside puddles or a wet tundish. Ask the owner if they have noticed any discharge.
4. Manually check the **temperature** relief valve. Lift the gear or twist the top on the integral test device on the relief valve for about 30 seconds to remove any residue that may have collected on the valve seat. Check that the valve reseats, discharges water and then re-seals.

Image 76a **Test the pressure and temperature and pressure relief valves**

It is common for the temperature relief valve not to reseat and a replacement might be required.

5. Manually check the **expansion** relief valve: Lift the gear or twist the top on the integral test device on the relief valve for about 30 seconds to remove any residue that may have collected on the valve seat. Check that the valve reseats, discharges water and then re-seals.

Image 76b **Check for clear flow through the tundish**

6. Check the discharge pipes from both the expansion valve and the temperature relief valve for obstruction. Check that the termination point has not been obstructed.
7. Check that D1, D2 and the termination comply with the Building Regulations.
8. Check the cylinder's operating thermostat setting (e.g. 60-65°C).
9. Isolate the gas and electrical supplies to the heating appliance, turn off the water supply and release water pressure by opening at least one hot tap.
    a. Drain inlet pipework where necessary to check, clean or replace the line strainer filter (on newer systems this is normally part of the pressure reducing valve assembly).
    b. Check the pressure in the expansion vessel and top up as necessary whilst the system is empty or uncharged. Use a tyre gauge and pump to recharge to operating pressure.

    With an older system, it is wise to check whether the expansion vessel diaphragm is intact by checking the pressure in the expansion vessel when the system is full of water. If the diaphragm is faulty, water will be released out of the Schraeder valve on top of the expansion vessel.

    If the cylinder incorporates an air gap as the expansion facility, reinstate the air gap as part of the procedure in line with manufacturer's instructions. Even though the bubble may still be intact within the cylinder, air is naturally absorbed by water and the gap can be reduced so this is the annual opportunity to replenish with fresh atmospheric pressure.

Image 76c **Check and clean the line strainer**

10. Reinstate water, electricity and gas supply. Run the system up to temperature and ensure that the control thermostats are working effectively and that the relief valves are not discharging water.
    a. Check incoming mains water pressure and flow rate upstream of the pressure reducing valve as a reference.
    b. Check the pressure, flow and temperature at the owner/occupier's outlet and compare against the mains supply reference above.

    This will identify any faulty components. It is suggested these tests are taken at the tap the furthest away from the unvented cylinder, as 55°C is a minimum requirement at any hot tap.

11. Complete any maintenance records and Benchmark forms.

Never make a temporary repair on an unvented system by leaving a valve out of the system or wiring out a thermostat. Always replace the parts 'like for like' and if you need to obtain replacement temporarily leave the cylinder out of service. Warranties will be rendered invalid if incorrect parts are installed. Also, people's lives might be at risk.

Any incorrectly installed systems should be reported to the owner/occupier and, if necessary, followed up by a report to the local Building Control Officer.

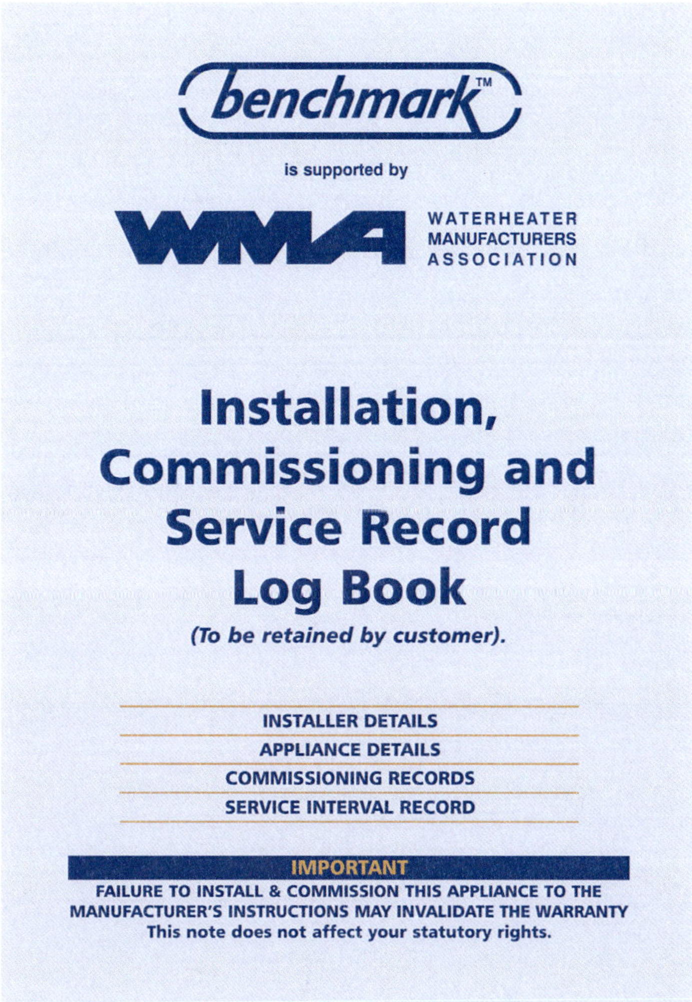
benchmark™

is supported by

WMA WATERHEATER MANUFACTURERS ASSOCIATION

Installation, Commissioning and Service Record Log Book

(To be retained by customer).

INSTALLER DETAILS
APPLIANCE DETAILS
COMMISSIONING RECORDS
SERVICE INTERVAL RECORD

IMPORTANT
FAILURE TO INSTALL & COMMISSION THIS APPLIANCE TO THE MANUFACTURER'S INSTRUCTIONS MAY INVALIDATE THE WARRANTY
This note does not affect your statutory rights.

Don't forget to complete the benchmark log book after installation, commissioning and servicing.

Diagnosing faults in unvented hot water storage systems

Faults are relatively rare in unvented hot water storage systems, the manufacturer's installation instructions will contain quite detailed procedures for identifying faults for their specific unvented hot water storage systems and these should be used as a key reference when diagnosing faults.

The following flow diagrams of key faults on a system incorporating an expansion vessel as the expansion device have been included as an example of the information provided to diagnose faults.

Fault A: Water discharge from expansion valve

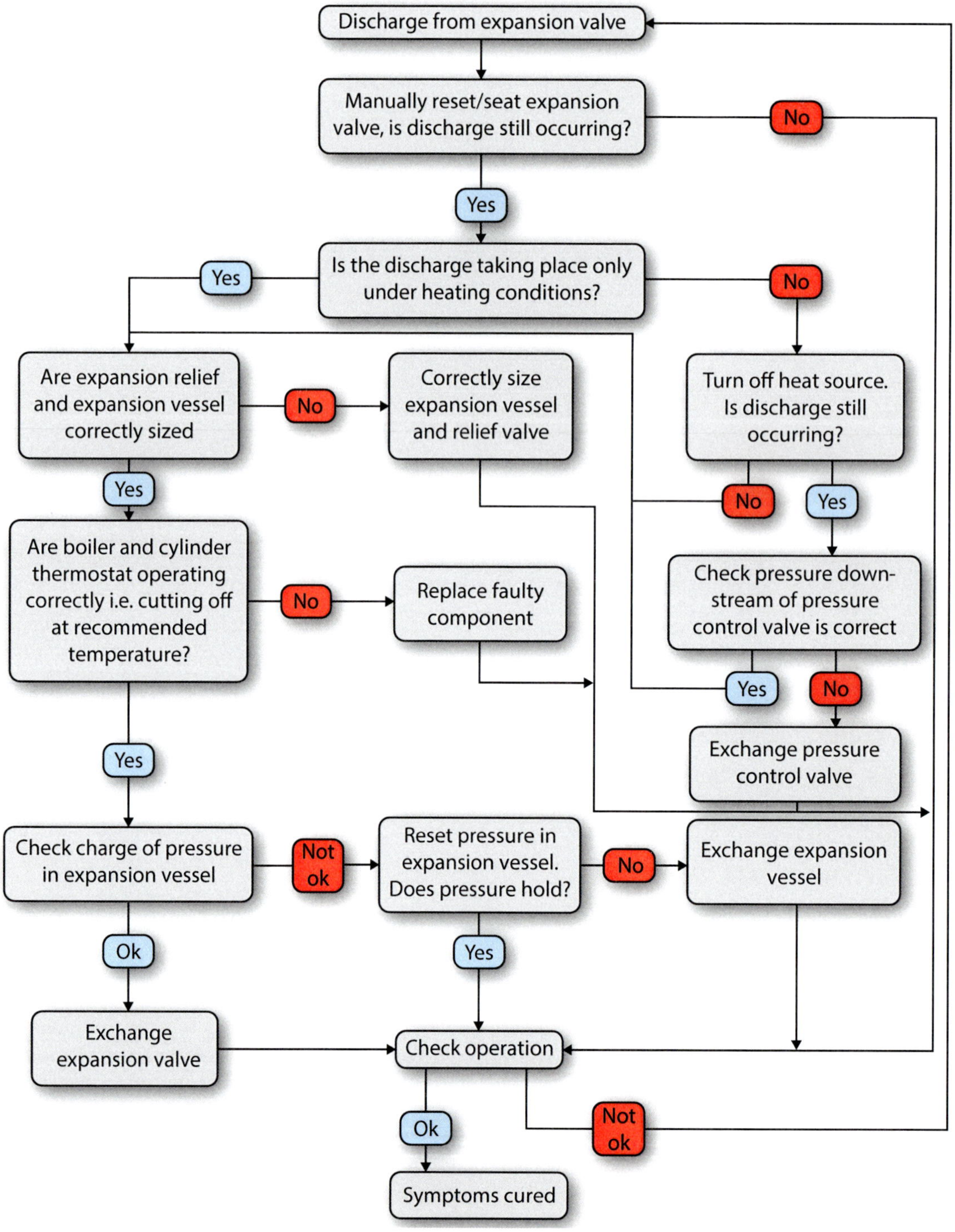

Internal expansion type cylinder

With bubble tops or cylinders incorporating an air gap as the expansion device, intermittent discharge from the expansion relief valve will in most cases mean that the air gap or bubble has reduced in volume. A simple solution to this problem is to:

- Turn off the water supply to the cylinder
- Open the lowest hot water tap in the system
- Hold open the temperature relief valve until water ceases to run from the cylinder
- Re-fill the system by turning on the mains supply. The air volume will automatically recharge as the bubble refills
- Turn off the hot tap when a smooth flow is seen

There are a number of manufacturers that incorporate internal bubbles in the cylinder and the principle of recharging the air is basically the same for all types. One brand has an internal floating baffle disc to separate the water and the air, intended to prolong the life of the air.

Other brands may reduce the pressure of the incoming water to control the internal water pressure, which in turn will also prolong the life of the air.

Always check the manufacturer's label, which is normally located on the cylinder next to the temperature relief valve; this will show the manufacturer's instructions for recharging the cylinder.

The air gap in the internal bubble type has disappeared

The main reason why the air gap in the internal bubble will disappear is that the oxygen in the air is absorbed into the water (H_2O). The time it will take for the air to be absorbed will depend upon the amount of oxygen already in the water.

The Water Regulations requires that neither type of expansion facility waste water, so periodic maintenance checks are required.

Fault B: Water discharge from temperature relief valve

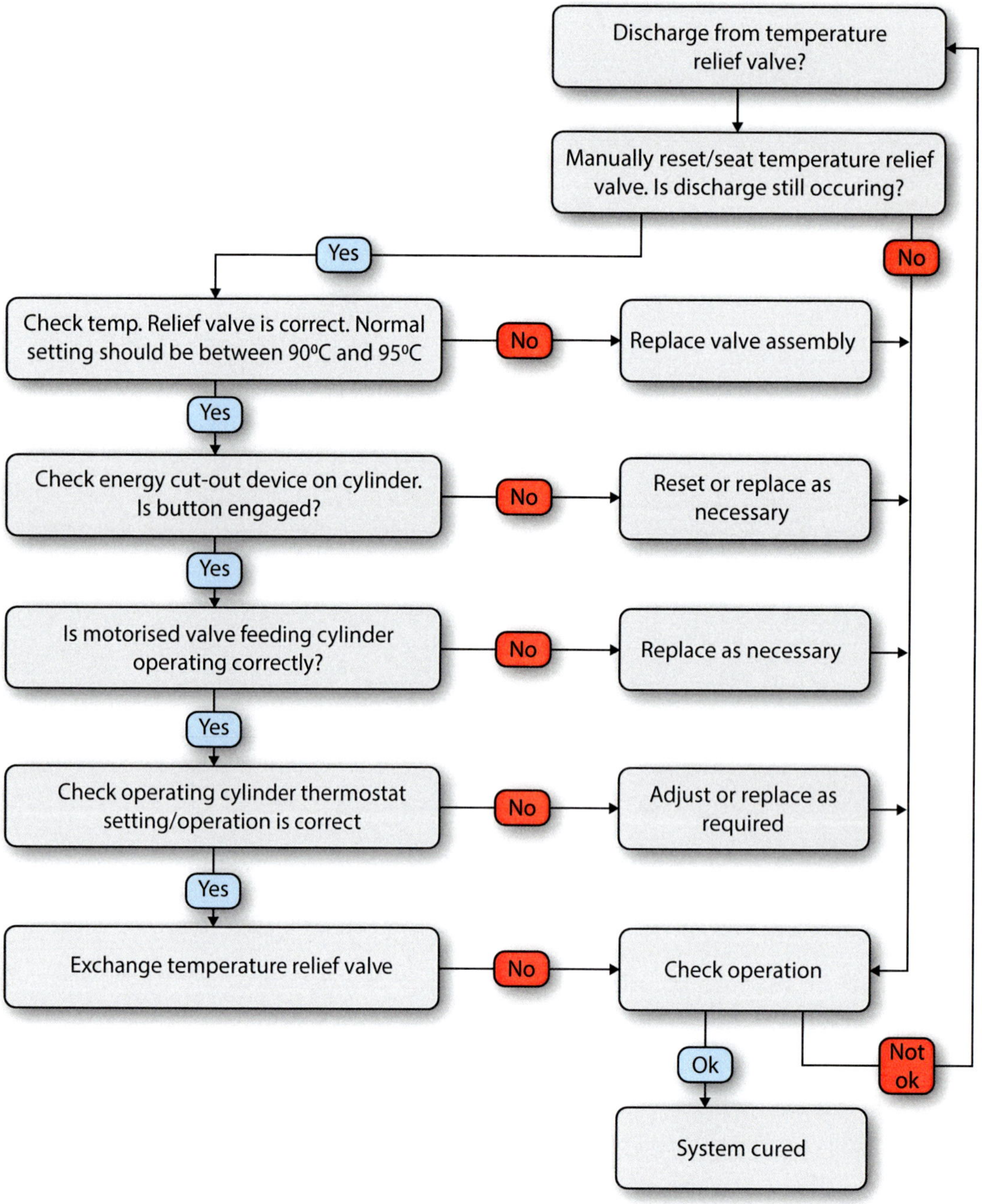

Fault C: Deteriorating flow from terminal fittings (taps)

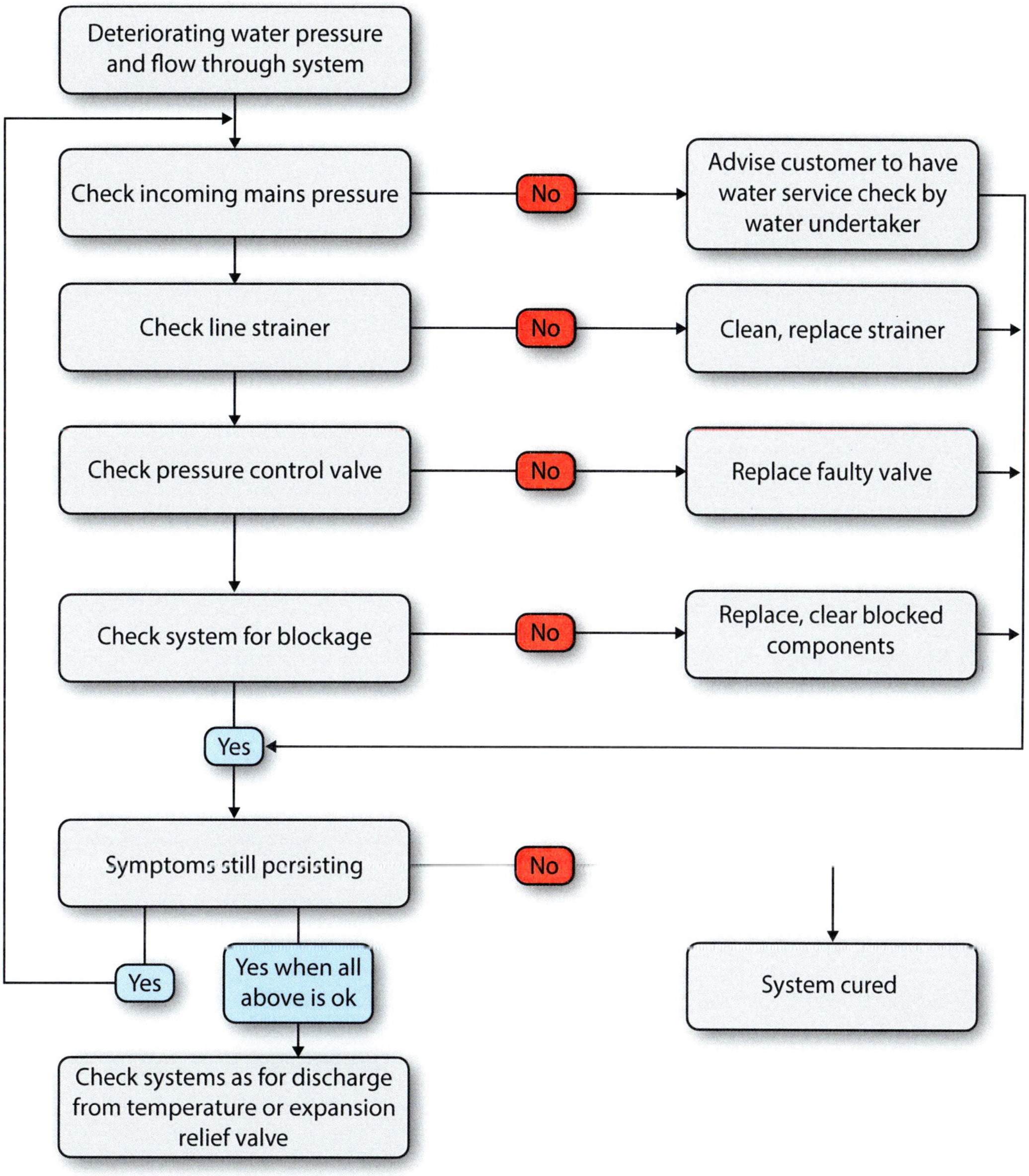

Vented system maintenance requirements

Here's a typical maintenance schedule for a vented hot water system.

Component	Recommended time interval for inspection	Procedure
Vented hot water system	12 months	General inspection of system.
Control valves	12 months	Operate to check that they close tightly and operate smoothly. Repair and renew as necessary. Valves should be clearly labelled on larger installations and the customer should be made aware of isolation valves.
Storage cisterns	6 months	Inspect for cleanliness and clean out as required. Make sure the lid is securely fixed and look for signs of leakage (or corrosion on older galvanized cisterns). Check linked cisterns for any signs of stagnation. Make sure that bearers and cistern base is sound and adequate. Check overflow pipes for any signs of blockage and for correct fall. Make sure that insulation is adequate and intact. Upgrade if necessary.
Float operated valves	6 months	Check operation and shut off. Adjust for correct water level. Re-washer/replace if required.
Terminal fittings	6 to 12 months	Re-washer, reseat or renew taps as required to prevent leakage. Tighten packing glands, check spindles for wear and correct operation. Clean spray heads on taps, showers and shower mixers.
Pipework	2 months	Check for any signs of leakage or corrosion. Tighten any loose fittings, joints, or components. Replace any vulnerable sections of pipework and fittings.
Thermostats	6 months	Make sure that boiler/cylinder/immersion heater thermostats/cut outs are working to specification. Check that cylinder temperature and temperature at the terminal fittings conforms to regulations.
Earthing and bonding	At time of other inspection	Make sure that existing earthing and bonding is intact, particularly following any pipework replacement. Get electrician to check the installation If in doubt about the condition of the earthing and bonding.

Now a few questions for you!

Self assessment questions

1. You have been called to an unvented hot water storage system that you have not installed or serviced previously and you find that the temperature relief valve port has been blanked and the valve is missing. You do not have a spare, what do you do?

2. You have been called to attend an unvented hot water storage system incorporating an expansion vessel in which the expansion valve is discharging water, list 3 possible causes.

   a.

   b.

   c.

3. You are asked to attend to an unvented hot water storage system incorporating an integral air bubble in which the expansion valve is intermittently dripping water.

   a. What is the most likely cause?

   b. What remedial action is necessary?

Self assessment answers

1. You should have identified that you would isolate the system and leave in an inoperable condition until such time as you obtained a new temperature relief valve, fitted and commissioned it.

2. You could have had one of the following and more besides!

   a) Vessel charge pressure too low

   b) Defective diaphragm in expansion vessel

   c) Defective pressure reducing valve

3.

   a) The air gap in the cylinder has reduced in volume

   b) Isolate the water supply to the cylinder, drain down through the temperature relief valve, turn on the water supply and fill the system, the cylinder air gap volume should be reinstated.

That's it for this module!

Module 4

Other Applicable Building Regulations

Contents

Aim of the module

On completion of this module and following attendance on the BPEC Vented and Unvented Hot Water Systems Training Course, you should be able to:

- State the other Building Regulations and Approved Documents that apply to the installation of hot water storage systems
- State briefly what effect these regulations and documents have on the installation of hot water storage systems

Introduction

Following the introduction of Part L of the Building Regulations in April 2002, tighter controls were imposed on 'controlled services' such as hot water storage systems and central heating. Part L has been updated a number of times since its introduction and each update generally requires a greater standard of energy efficiency. The current versions are 2010 but have been amended in 2013.

This means that there are specific minimum standards that apply to their installation, commissioning and control systems. Here we will explain what those minimum standards in relation to hot water systems. Because G3 covers vented and unvented systems we'll cover the requirements for both.

In January 2005 a new Part P of the Building Regulations was introduced which details the competence requirements for enterprises (such as an individual/sole trader) that install electrical circuits and equipment as part of another overall task e.g. someone installing central heating and/or a replacement vented or unvented hot water cylinder and its controls.

Although reference is made here to Part L and Part P of the England and Wales Building Regulations, the requirements are the same for the devolved administrations. Refer to the table on page 8 for which legislation is applicable in Scotland and Northern Ireland.

Part L and hot water storage systems

Part L of the Building Regulations was introduced in April 2002 and amended in 2006, 2010 and 2013. It is a legal requirement that a plumber ensures that he/she complies with the requirements of the approved documents. There are two parts:

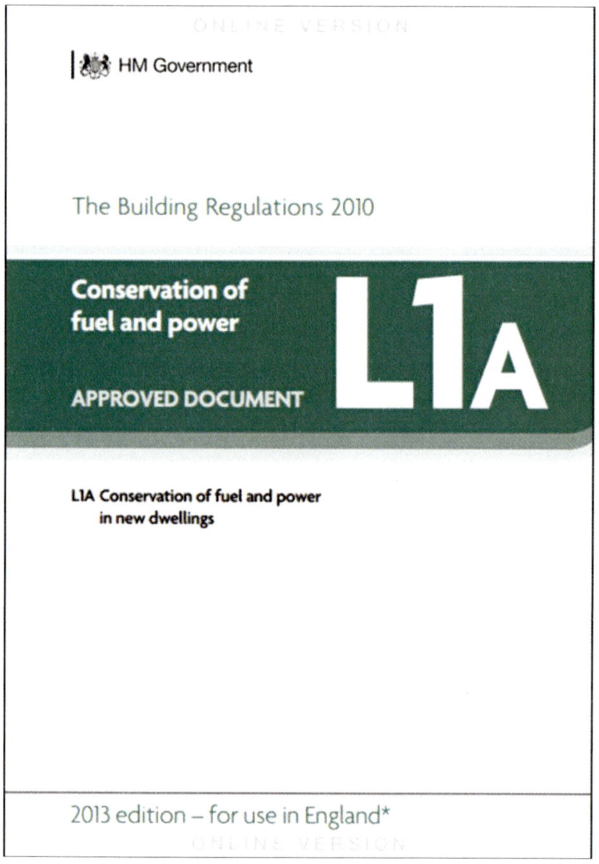

Image 77a L1A – new dwellings

Image 77b L1B – existing dwellings

The Regulations lay down stringent requirements for work carried out in new build properties or when work is carried out in existing properties. Let's take a brief look at what the L1 documents require.

For existing dwellings the regulations relating to hot water and heating cover:

- Replacement heating systems (including boilers and vented and unvented hot water storage cylinders)
- Heating and hot water controls
- Commissioning of heating and hot water systems and the provision of instructions for users

The regulations also deal specifically with hot water systems in terms of:

- Standards for vented hot water vessels conforming to BS 1566 and BS 3198
- Standards for unvented cylinders conforming to BS EN 12897:2006 or be approved by the BBA, WRc or KIWA
- Temperature control
- Insulation of pipework
- Alternative approaches for hot water systems
- Commissioning of hot water system, in compliance with the Code of Practice Benchmark 30
- Energy Related Products performance labelling

For new dwellings the regulations include certification of hot water and heating systems to show that they've been correctly installed and commissioned, and that operating instructions have been left for the user.

For completely new systems, the recommended minimum system design requires the installer to:

- Calculate the boiler output
- Select a boiler on the basis of SEDBUK rating
  - Although still referred to by many as SEDBUK the boiler efficiency is now found on the Product Characteristics Database at http://www.ncm-pcdb.org.uk/sap/searchpod.jsp?id=17
- Use a fully pumped system for both the space heating and water heating
- Install controls for wet systems to include:
  - Room thermostat
  - Two-channel programmer with standard thermostat or hot water time switch with programmable room thermostat
  - Motorised valves for pumped space and water heating
  - Boiler interlock (this is the wiring arrangement of the boiler and controls and should be wired so the boiler can't fire unless there's a demand from either the space heating or domestic hot water)
- Certify heating and hot water systems to show that they've been correctly installed and commissioned, and that operating instructions have been left for the user

One of the major requirements of the legislation as far as we are concerned is: "The certification of heating and hot water systems to show that they've been correctly installed and commissioned and the provision of operating and maintenance instructions for users". Remember the Benchmark log book!

In order to be approved for installation in the UK, unvented cylinders must comply with the requirements of BS EN 12897:2006. Basic cylinders must be factory insulated to a level where the standing heat losses in 24 hours will not exceed 1.6 x (0.2 + 0.051V) kWh, where V is the cylinder capacity in litres.

What this complicated formula equates to in real terms is about 0.8 watts per litre for an average 117 litre cylinder.

High performance cylinders have a formula of 1.28 x (0.2 + $0.051V_{2/3}$) kWh, which equates to 0.64 watts per litre for a 117 litre cylinder.

Ultimately, cylinders must be very well insulated, a simple sectional cylinder jacket is no longer sufficient.

Part L also lays down requirements for the recovery times for cylinders and unvented cylinders must comply with BS EN 12897:2006. BS EN 12897:2006 states that for a 120 litre cylinder with a primary flow rate of 15 litres/minute, the cylinder must reheat in a maximum of 20 minutes.

120 litre cylinders with primary flow rate of 20 litres/minute must reheat in no more than 17.5 minutes.

Part L also requires that time and temperature controls are fitted to heating and hot water storage systems to prevent any unnecessary waste of energy.

In reality all unvented cylinders will be supplied with built in temperature controls. In the case of indirect cylinders the cylinder thermostat and the energy cut-off device must be wired in series to ensure that the over temperature protection is maintained.

Domestic Heating Compliance Guide

Image 78 The Domestic Heating Compliance Guide

The Domestic Heating Compliance Guide is a second tier documents that is referred to in L1A and L1B as a source of guidance to help to comply with the L1 regulations for space heating and hot water systems.

The guide was prepared in consultation with the industry bodies. It covers conventional means of providing primary and secondary space heating and domestic hot water in dwellings.

Status of the compliance guide

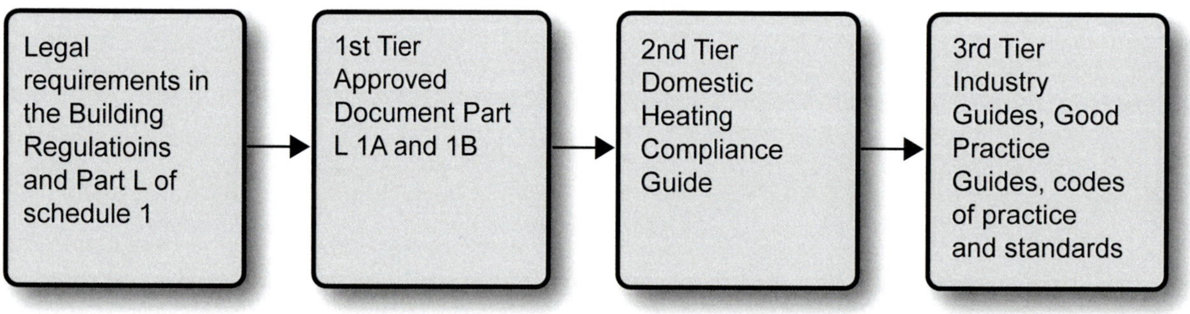

Other general requirements related to all domestic hot water systems covered in the guide includes:

- All "hot pipes" connected to the hot water cylinder (this includes the primary flow and return connections and draw off/vent pipe) should be insulated for at least 1 metre in length or up to the point where they become concealed (as in L1)
- When a cylinder is being replaced in a semi gravity system which has no temperature control, the Compliance Guide stipulates that as a minimum, a thermo-mechanical control valve should be fitted to regulate the circulation through the cylinder.

Commissioning

Under the requirements of Part L the installer must complete a full and accurate commissioning of the controlled service in question. The process must include the completion and issuing of a commissioning certificate. Benchmark certificates will meet this requirement.

A copy of the certificate must be provided to the local building control body and to the customer. In many cases the customer copy takes the form an adhesive label to be fixed to the cylinder by the commissioning engineer.

Part L certification

Part L also details the minimum and best practice requirements for new and replacement heating systems for gas, LPG and oil, the minimum seasonal efficiency requirements for boilers and the minimum and best practice for heating controls.

To go into detail here would take far too long as this is a 1 day training programme.

The best way that you can prove that you can comply with the requirements of Part L is to undertake the BPEC Part L Certificate in Energy Efficiency. Ask your course tutor for details or telephone 0845 644 6558 or www.bpec.org.uk

Part P

Part P of the Building Regulations was amended in January 2005 and again in 2013 to allow competent persons, who are members of appropriate registration schemes, to install electrical circuits and controls as part of an overall task in domestic premises.

For example someone installing a central heating system and controls can install the electrical cables and components, provided they hold an appropriate qualification and can issue valid self-certification certificates. They must also notify the local building control body within 30 days that the work has been completed.

Some minor electrical work, such as changing a room thermostat or a central heating programmer which does not involve any changes to the fixed wiring is not notifiable under Part P but the individual must still be competent to carry out the work.

Generally all work which includes alterations or additions to the fixed wiring of a domestic premise in 'special areas' such as kitchens, bathrooms, en-suites, shower rooms, utility rooms or outdoors will be notifiable. It is likely that an unvented cylinder may be installed in most of these 'special areas'.

Part P commissioning

Under the requirements of Part P the installer must complete a full and accurate inspection and testing of the electrical work carried out as well as the existing installation.

Depending upon the type of work carried out, the process must include the completion and issuing of a schedule of test results, a minor works certificate or a full electrical installation test certificate, which must be left with the customer.

If the work is notifiable, then the local building control body must also be notified within 30 days that the work has been completed.

Part P certification

To fully explain the requirements of Part P here would require a complete new volume but in simple terms, if you carry out electrical work that involves extensions or alterations to the fixed wiring of a domestic dwelling you need to hold a certificate of competence and to be registered with an appropriate Part P registration body.

The best way to prove that you can comply with the requirements of Part P is to undertake the BPEC Part P Defined or Full scope electrical training and assessment courses. Ask your course tutor for details or telephone 0845 644 6558 or visit www.bpec.org.uk

Electrical work in Scotland

All electrical work on the fixed wiring of a dwelling in Scotland must be carried out by a qualified electrician. Suitably trained and competent engineers can carry out work on the 'appliance'.

The appliance can be anything after the isolation switch (fused spur) for example, the controls of a central heating system, a hot water cylinder or shower.

The best way to prove that you can comply with the requirements for electrical work in Scotland is to undertake the BPEC Electricity for Plumbers electrical training and assessment courses. Ask your course tutor for details or telephone 0845 644 6558 or visit www.bpec.org.uk

Self assessment questions

1. When was the new Part L of the Building Regulations first introduced?

2. When was the new Part P of the Building Regulations first introduced?

3. Part L requires a minimum level of insulation for all cylinders. In addition to the cylinder what else must be insulated in the cylinder installation?

4. Under Part L what is the maximum recovery time for a 120 litre cylinder with a primary flow rate of 20 litres/minute?

5. In order to avoid unnecessary waste of energy, what must be fitted to heating and hot water systems?

6. In order to carry out electrical work on the fixed electrical wiring to an unvented cylinder in a bathroom, what 2 things must you do?

Self assessment answers

1. April 2002
2. January 2005
3. All pipework within 1 metre of the cylinder
4. 17.5 minutes
5. Time and temperature controls
6. A Part P certificate of competence and registration with a Part P registration body